my brain MADE ME DO IT

Eliezer J. Sternberg

my brain MADE ME DO IT

THE RISE OF NEUROSCIENCE *and the* THREAT TO MORAL RESPONSIBILITY

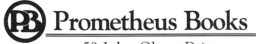
Prometheus Books

59 John Glenn Drive
Amherst, New York 14228–2119

Published 2010 by Prometheus Books

Inquiries should be addressed to
Prometheus Books
59 John Glenn Drive
Amherst, New York 14228–2119
VOICE: 716–691–0133
FAX: 716–691–0137
WWW.PROMETHEUSBOOKS.COM

14 13 12 11 10   5 4 3 2

Library of Congress Cataloging-in-Publication Data

Sternberg, Eliezer J.
    My brain made me do it : the rise of neuroscience and the threat to moral responsibility / Eliezer J. Sternberg.
        p. cm.
    Includes bibliographical references and index.
    ISBN 978–1–61614–165–3 (pk. : alk. paper)
    1. Free will and determinism. 2. Responsibility. 3. Ethics. 4. Neurosciences.
I. Title.

BJ1461.S695 2010
123'.5—dc22

                                                                2009047762

Printed in the United States of America on acid-free paper

For Danny, Benny, and Rebecca

CONTENTS

FOREWORD

My introduction to the brain sciences came almost fifty years ago when, as a youngster, I came across a Ripley's Believe It or Not cartoon about Phineas Gage, an unfortunate railroad worker who, one hundred years earlier, survived an accident in which a metal rod was driven through his cheek and out of the top of his skull. What stuck in my memory—along with other Ripley's unforgettables such as the man who had the hiccups for seven years and the monk whose nails were three feet long—was that although Gage was blinded in one eye and his temperament changed, *he lived another twelve years in good health!* Ripley's provided the bold italics here to make sure I got the believe-it-or-not element and underplayed the neurological and personal effects of Gage's accident.

I was reminded of this first encounter because Sternberg retells the story of Phineas Gage. But now, fifty years later, we understand not only why he survived but also a fair amount about the cognitive and attentional processes that lay behind his personality changes, as well as the way those processes are localized in his brain anatomy. The sense of wonder is still there—Phineas Gage's story continues to capture the imaginations of visitors to the various Ripley museums—but we now understand *why* what happened to *his brain* affected *him* in the way it did.

It is part of our nature to wonder, to pay a special kind of attention to unexpected phenomena that we don't understand. But because there is at least a rough correlation between the extent to which our ancestors understood their world and their success in surviving its challenges and reproducing, it is also part of our nature to try to move beyond wonder and to seek understanding. In our modern scientific culture, this quest for understanding has moved well beyond the demands of practical action. It is now a mile wide *and* a mile deep: we want to understand everything in the world around us, and we want that understanding to get to the very bottom of things. And we want that understanding to have a special character: we want *scientific theories.*

Once we think about the scope of these ambitions, it becomes plain that they are not limited to the world *around* us. We also want to understand ourselves. We, after all, are a part of the world. We, too, have a nature that's a mile deep and waiting to be probed. Our scientific ambitions cannot be satisfied unless we include our *selves*—our physical natures as well as our minds—in the project.

We have learned a great deal over the past century about our physical nature; the mind has proved more elusive. Is it distinct from the body? Perhaps just a different way of talking about the body? After a number of false starts, and the development of new technologies and approaches, we have begun to make real progress in understanding the mind, and there is an emerging consensus that much of that progress involves our growing understanding of the brain. We are still in the early stages, but the consensus has it that we are finally on the right track, and that if we understand the brain, we will understand the mind.

But despite the consensus, and the undeniable progress in our understanding of the brain, some have worried that we are headed into dangerous waters. What if the understanding of the world that our science builds challenges our commonsense conception of that world? This problem is an inherent feature of our scientific gestalt: once we set ourselves the task of getting behind *appearances* and understanding the underlying *reality*, we have already conceded that our prescientific conception is just appearance after all. The major shock came at the very outset of our scientific era: we had to learn to think of natural phenomena as all governed by impersonal laws and not the intent of an imminent purposive God. Rainbows are nothing but light and water droplets, the Earth is just one planet among countless others, and the stars are nothing but fireballs. A century later, the theory of evolution provided another shock to the system: living things were not created, but evolved by random mutation and natural selection. As a culture we have (for the most part) absorbed these shocks and made our peace with the new scientific picture of the physical world around us. We have learned to "stand corrected" and to defer to science. Even when we can't "stand corrected" because we can't fully grasp the "correct" answers (here I'm thinking of advances in particle physics like quantum mechanics and string theory), we still defer.

But now it's not the world *around* us that we need to rethink; it's *us*. Are we ready to defer when the subject matter is *we ourselves* and the world *inside* us? On most matters, what the brain sciences tell us about ourselves can be absorbed without much difficulty. There is no problem here. We can comfortably accept recent findings about the way our body is represented in the brain, and even adjust to the fact that pain and sensation does not *really* occur in our bodies in the way we intuitively think (even though it continues to *appear* to do so no matter how much science we know). We can even revise our commonsense understanding and learn to accept that most of the cognitive processing that makes up our thought happens below the level of conscious awareness. These are the easy adjustments.

But our scientific understanding of the brain is beginning to strike deeper and may call for more difficult adjustments. Some neuroscientists tell us that what we are—our beliefs, moods, desires, motivations, and so on—are all features of our brains; that these features were caused by prior events over which we have no control; that our actions are brought about by changes in the environment and changes in our brains over which we have no control; that we do not in any real sense act *freely*; that *we* are not what *we* think *we* are and that our conception of ourselves as being the *authors* of our *freely chosen* actions is not part of the ultimate reality of things; that it is all just *appearance*; and that the baggage that was carried by this appearance—specifically, the idea that we are morally responsible for what we do and that our actions did not *just happen*—needs to be abandoned too. Can we defer to science here too? Can *we* defer?

These new "shocks" that I've portrayed as the legacy of the modern scientific assault on the brain actually have deeper roots—the tension has in one sense been there all along. We all naturally take ourselves to be freely authoring our actions. But many of us are puzzled (at some point in our lives) about how this can be if we, and the choices we make, are in the end the products of causes beyond our control. How can we be held morally responsible for anything we do if all our actions are inevitable? The puzzlement is in this sense independent of any particular findings about the brain. All it needs to get off the ground is the idea of the brain and the body as deterministic, cause-and-effect driven, biological systems.

Most of us let go of the puzzle and get on with our lives; we learn to leave these concepts all knotted up just as we found them. Those who can't or won't let go become philosophers. Some philosophers accept the scientific image and hold that free will is an illusion. A close variant: free will as an escape from causal determinism *is* an illusion, but in another conceptualization it is really compatible with causal determinism. Those who resist the scientific image have harder choices. Some—a smaller and smaller minority these days—have second thoughts about the assumption that humans are nothing more than biological systems: perhaps we have souls/minds that are *not* physical, and our freedom is connected to the role of the soul/mind in our actions. Yet another view is that there are mysteries about the human mind and human action that most likely outstrip our intellectual faculties: we would be better off if we devoted ourselves to problems that we *can* solve and let this sleeping dog lie. Other philosophers, with an interest but no firm stance, devote themselves to untangling many of the knots, to making distinctions and introducing refinements in the way the original puzzlement is formulated, and to providing maps that mark out the nuanced positions one might take with regard to these puzzles. But few think we have truly solved the original puzzle. For those who would only be satisfied by a real solution, the impulse to set aside the problem and to get on with life seems a healthy strategy.

The effect of scientific research in the brain sciences is that it makes it harder to set aside the problem. As the neurosciences uncover more and more details about the minute changes in our brains that underlie our actions, things are brought to a head. The biological approach to human nature has worked too well. It is harder to hold on to the idea that there is a nonbiological soul/mind at work in our lives. The view that human action is an ultimately incomprehensible mystery begins to seem petulant as more and more that was once mysterious becomes clear.

There is another factor. It is not easy for some of us to accommodate the general idea that we are causally determined. But for many of us, the shock is easier to absorb if this causal determination happens "in the dark": if it is very, very, complicated and our behavior is practically unpredictable. But the brain sciences are threatening to shine a light on these causal processes. As Sternberg's profiles and vignettes make plain, our intentions, our beliefs, and our thoughts are all under a bright spotlight. Their true nature, their role, and their power are all being questioned and rethought by the neurosciences. What was once part of the private theater of the mind may soon become a well-understood biological process, and it will be harder to ignore the challenge of our original puzzle: is our sense of ourselves as free to choose and to act a kind of illusion?

What's happened, then, is not that we've discovered causality and determinism in places where everyone assumed it was absent. It's rather that it has become harder to ignore the causality as the details begin to come clear. Sternberg is rightly perplexed about cases where, for instance, a defense attorney can point to irregularities in a defendant's brain chemistry that are clearly connected to impulsiveness and violent behavior. Now the problem becomes practical, not just philosophical. The juror is forced to ponder if the perpetrator of the heartless murders is really *responsible* for them given that the mechanisms in his brain that should have inhibited this impulsive violence—and that by our good fortune protect us from our most violent impulses—was dysfunctional. And what if we discover neurological differences between those who engage in Ponzi schemes and the rest of us? The fact is that we can't just leave it to philosophers to ponder these problems. We—as citizens, as physicians and therapists, as law enforcers, as parents and children, as jurors, as jurists, and so on—need to somehow draw the line about matters of personal and moral responsibility. But is there really a line to be drawn?

Sternberg's book is an excellent guide for those who want to confront these puzzles in their full scientific and philosophical complexity. It clearly explains many of the fascinating scientific advances in our understanding of the brain-behavior connection, and it carefully considers their relevance to the free will question. It is especially good at making these complicated theoretical issues come alive in vivid case studies and extended hypothetical examples.

But Sternberg wants to do more than just perturb us with the problem. He wants to show us the way out. As he sees it, neuroscientific reports of the demise of the self and free will are, as Mark Twain said, "greatly exaggerated." His central critique here is that neuroscientists who have tried to link deliberation and choice to brain activity have focused on only the simplest sorts of cases—like the decision to move a lever in one direction or another. This is as it should be, because the scientific project is to show how complexity is built out of simple elements and operations. But we should not just *assume* that the findings about simple choice behaviors will ultimately "scale up" to explain what happens when we reflect deeply on what he calls "boundless" problems. These are problems for which the alternatives, and the relevant factors to be considered and weighed, cannot be specified antecedently. Difficult life choices are paradigm cases of such boundless problems. From the inside, deliberating about these choices feels like roaming freely in our unconstrained mental landscape, with nothing fenced off or out of bounds. This ability to reflect in this "anything goes" way, according to Sternberg, is what sets humans apart and is the true expression of our free will.

Our sciences, despite advertisements to the contrary, have no theory that can come to grips with this sort of deliberation. And Sternberg thinks that they never will, because such thinking is not the result of deterministic processes; it is not mechanical or algorithmic. If he is right, then *we*, as deliberating agents, can escape the clutches of neurobiological determinism with our free will unscathed. *We* should not surrender prematurely; *we* need to be brave and stand fast.

Is Sternberg right? Can *we* really do what no deterministic biological system can do? Read the book and decide for your*self.*

Jerry Samet
Associate Professor of Philosophy
Brandeis University

ACKNOWLEDGMENTS

I feel a moral responsibility to convey my deepest thanks to those who have helped bring this book into fruition. I am indebted to John Lisman, Carol Palm, Jerry Samet, Robert Sylwester, and Andreas Teuber for their guidance throughout this process, and for being wonderful mentors for me over the past few years. I want to thank Sharona Hakimi for helping me with the index. Thanks also to Melanie Braverman, Norman Doidge, Jeremy Heyman, Eli Hirsch, Joseph LeDoux, Ro'ee Gilron, Daniel Millenson, Sam Packer, Julie Seeger, and Marion Smiley for assisting me with revisions, providing me with useful comments, or otherwise giving me the help I needed whenever I asked for advice.

The writing and preparing of the manuscript for this book was certainly a demanding process, and I am grateful for the invariable support and encouragement of my friends Josh Balderman, Perry Bell, Avi Cooper, Anton Eriera, Jon Freed, Chippy Hait, Charlie Gandelman, Greg Goodman, Sharona Hakimi, Noah Kaplan, Marnina Koschitzky, Ryan Schwab, Michael Sherman, Michael Shoretz, and Sara Smith.

Finally, I want to thank my family. My siblings, Daniel, Benjamin, and Rebecca, have always been my most dedicated fans and closest allies, and my parents, Ernest and Zohara, are probably the only people who are more emotionally invested in my work than I am. It is my blessing to have grown up with them and to have been a beneficiary of their love, care, and endless compassion.

INTRODUCTION

In his 1864 book, *Notes from Underground*, Fyodor Dostoevsky tells the story of a man, whose name we never know, who is plagued by his own consciousness. He thinks of himself as having a sickness of consciousness that leaves him paralyzed with indecisiveness. He believes he is *too* conscious, overly aware of his own thoughts and intentions, and excessively self-critical. To make matters worse, he is paranoid that his decisions are being controlled by the expectations of society, and even by the mechanical forces of his body. Written as a memoir, the book expresses the Underground Man's endless frustration as he tries to exert his will upon the world, struggling to act in such a way as to convince himself that he is in control of his own thoughts and actions.

He intentionally bumps someone walking toward him to demonstrate that he didn't want to step out of his way. He claims to enjoy the pain of a toothache because people expect him not to. He attends a dinner party of old schoolmates whom he secretly would like to befriend and decides to openly insult them. Though he realizes that the other guests don't want him there, he paces around in indecision as he contemplates revenge but refuses to leave. Every time he senses that others might have a certain expectation of how he will behave, he acts differently just to spite them.

One day, the Underground Man meets a prostitute named Liza who engages him in conversation. For no particular reason, he decides to lecture her about the indignity of her profession. The Underground Man is surprised to find that Liza is affected by what he says, and he is excited by the prospect of being able to exert influence over her. Liza responds to his bitter cynicism with patience, and over time the Underground Man begins to develop feelings for her. However, these feelings anger him. They add a frustrating complexity to their interaction, and the Underground Man is wary of how they might affect his behavior. Therefore, he mocks her unceasingly until she leaves him alone to sit with his thoughts.

Nevertheless, Liza remains tolerant of his behavior and comes to visit him. The Underground Man realizes that she might be the only person who views him affectionately, the only person who might one day understand him. It is clear to him that he should invest in this new relationship. But being that this is the decision he *ought* to make—the decision any sensible person in his position would make—the Underground Man decides to be spiteful and act differently. He throws money at Liza as a sign of disrespect, humiliating her and causing her to leave for the last time. The Underground Man returns to his inner mental battle, and to his cynical ramblings.

The epigraph that begins this book is taken from one of the Underground Man's notes.[1] Throughout Dostoevsky's novel, the Underground Man grapples with the question of whether he is truly in control of his actions, always confronted with the possibility that he is merely one insignificant component (an "organ stop") in a purely mechanistic world (the organ). What's more, he fears that someone with enough information about him could potentially create some kind of formula with which to predict his every decision. It is for this reason that he tries to act erratically, hoping to evade the clutches of inevitability. He strives only to show that he has the ability to make his own choices, the power to make an impact on his environment—some bit of humanity to give meaning to his existence.

The problem the Underground Man poses is even more relevant today than it was in 1864. In fact, it is something that has bothered me for a long time. With the development of the field of neuroscience, we understand more and more about how human behavior is related to the interactions of neurons and their associated chemicals in the brain. Most modern scientists now work with the assumption that the mechanics of those biochemical entities is the sole determinant of human behavior. As we know, bio-chemical interactions follow a strict set of laws, the laws of chemistry and physics, to the point where they could be represented in the form of mathematical equations. If the thoughts we have and the choices we make are caused by the chemicals in our brains, then they must be determined by the laws of physics as well, so in theory they, too, could be represented as mathematical equations.

This is exactly the possibility that terrifies the Underground Man. Without freedom of choice, he cannot hold people responsible for their actions (whether the consequences are good or bad), because they *didn't have a choice*. Without freedom of choice, the Underground Man cannot claim to have an impact on the world, because he is just a pawn of the natural world, controlled by its rules. That is what leads him to be spiteful, to act in a way he feels no one could predict.

If it is true that biological mechanisms control everything we do, where does that leave us? It means we are machines; we are organ stops.

As for me, I really don't want to be an organ stop, and this feeling motivates me to search for a way out of this quandary. Though the threat of neuroscience to free will and moral responsibility is strong, there is a way they can be reconciled. There is a way that we can maintain both our trust in the methods and findings of brain science and our belief in the sanctity of the human being as a purposive, moral agent, or at least that is the argument I work toward here. In the chapters that follow, I will make the case that such reconciliation is plausible and offer an idea of how these views might be connected.

This book can be broadly divided into three sections. The first (chapters 1–4) explores the depth and the implications of the problem that I have just briefly introduced. The second section (chapters 5–13) outlines the fascinating developments in the neurosciences that seem to threaten our free will and moral responsibility. The third section (chapters 14–18) offers a possible resolution. As a whole, I intend this book not as simply a stage for me to present an extended line of reasoning, but as an exploration of fundamental questions of freedom, morality, and human purpose.

If some day they truly discover a formula for all our desires and caprices—that is, an explanation of what they depend upon, by what laws they arise, just how they develop, what they are aiming at in one case or another and so on, and so on, that is, a real mathematical formula—then, after all, man would most likely at once stop to feel desire, indeed, he will be certain to. *For who would want to choose by rule?* Besides, he will at once be transformed from a human being into an organ stop or something of that sort; for what is a man without desire, without free will and without choice, if not a stop in an organ?

—Fyodor Dostoevsky

1

THE NEFARIOUS NEURON

Just after midnight on February 17, 1991, Stephen Mobley walks into a Domino's Pizza in Oakwood, Georgia, holding a silver, semiautomatic pistol in his hand. Only the store manager, twenty-four-year-old John Collins, is inside. Mobley points the gun at Collins and orders him to open the cash register. Collins does as he is told. As Mobley stuffs his jacket with stacks of bills, the store manager retreats to the corner and sits, silently trembling.

Having emptied the cash register, Mobley pauses. He looks over at Collins, who is clutching his hair and on the verge of tears. He barks at the store manager to approach him, ordering him to get on his knees. Now sobbing, Collins obeys. Mobley walks behind his victim and places the weapon against the back of his sweat-soaked head. Collins pleads with him for mercy, begging him to take the money and go. Mobley pulls the trigger.

After being captured one month later, Mobley does not appear to regret the murder. "If that fat son-of-a-bitch had not started crying, I would never have shot him," he says. He tells one of his guards that he plans to apply for the night manager position at Domino's, since he knows there is an opening available. He points to another guard and claims that he is "beginning to look more and more like a Domino's Pizza boy everyday." He keeps an empty Domino's Pizza box with him in his cell.[1]

The task of defending Stephen Mobley is a difficult one. What can one say in favor of a killer who expresses no remorse? Mobley's attorneys begin with the medical examination, giving him a full physical and psychological workup. Their findings are scanty: Mobley does not have any physical, psychological, psychiatric, or neurological diseases. He doesn't have bipolar disorder, or schizophrenia, or Alzheimer's. The medical results return with only one minor discovery: Mobley has a slight deficiency of the enzyme monoamine oxidase A, responsible for breaking down serotonin, dopamine, and several other key neurotransmitters in the brain. The defense counsel knows that this is nowhere near enough to argue that Mobley is insane. Nevertheless, the team of lawyers decides to make a bold move. They declare that communication between neurons in Mobley's brain, influenced by the relative concentrations of neurotransmitters, caused Mobley to murder his victim. Mobley's attorneys release a statement:

> Stagnant MOMA [monoamine oxidase A] activity among affected males resulted in the excretion of abnormally high amounts of the neurotransmitters serotonin, norepinephrine, dopamine, and epinephrine, all of which are normally broken down in the body using MOMA.... When these neurotransmitters accumulate in abnormal amounts due to a defect on the MOMA gene, affected individuals will have trouble handling stressful situations, causing them to respond excessively and at times, violently.[2]

Effectively, their position is that Mobley cannot be held fully responsible for his crime because that crime was determined by his brain. His brain made him do it.

In the end, however, the jury was not convinced. Mobley was found guilty and subsequently executed on March 1, 2005.

Looking back on this case, what stands out to me is not the jury's verdict or the judge's sentence, but the defense counsel's strategy. Monoamine oxidase A is an enzyme that operates in every human brain, one of many that regulate the level of neurotransmitters and other brain chemicals, which in turn regulate communication between neurons. The amount of any given neurotransmitter in the brain, such as serotonin, is in a constant state of flux. It changes based on the needs of cells, the strength and number of incoming stimuli from the environment, and the frequency with which certain synapses are activated. The concentrations of chemicals in the brain vary from moment to moment and from person to person.

Are we to believe that one concentration of an enzyme causes a man to commit murder, another level of another enzyme causes him to lie, and another level of yet another enzyme causes him to give to charity?

Mobley's attorneys defended their client by claiming that his crime was caused by the interactions between neurons in his brain, through the exchange of various chemicals such as monoamine oxidase A. They chose this strategy because they assumed, as most people do, that the way a person behaves results from how his brain operates. But how is the case any different for someone with a perfectly healthy brain? The neurons in that person's brain are still firing, presumably generating his behavior.

Take, for example, a perfectly healthy person who tells a lie. What actually causes her to lie? She might provide a detailed rationale about her needs to achieve certain goals or her fear of revealing certain truths, but a scientific investigation will tell a very different story.

Deep within her brain, within the gray and white flesh of the cerebral hemispheres, through the weaves of neurons and glial cells, in between the delicate, arching tendrils of neighboring axons and dendrites, calcium ions are moving in and out of cells, regulating the exchange of tiny packets of the signal molecules known as *neurotransmitters*. These minute transactions create a signal that travels from the axonal bud of one neuron, down the snaking dendritic fiber of the next, propagating across to the axonal stem that follows, on and on from axon to dendrite and from neuron to neuron until finally the muscles are stimulated and the person speaks.

This woman is not mentally ill, any more than Mobley was as he stood with his gun to the head of his victim, contemplating his next move. Should he shoot Collins or just walk out the door?

When presented with a choice, I, too, may contemplate my next move, as in whether to lie or whether to tell the truth. I may take some time to consciously reflect on which option is best, based on my needs and values, but, at precisely the same time that I am supposedly steeped in reflection, billions of tiny calculations that I am unaware of are occurring in my brain—also resolving what I should do.

But if both these systems are at work, where does the decision come from? Do *I* control my decisions or does my brain?

Many neuroscientists today are convinced that it is my brain. The scientist Francis Crick writes that "'you,' your joys and your sorrows, your memories and your ambitions, your sense of personal identity and free will, are in fact no more than the behavior of a vast assembly of nerve cells and their associated molecules."[3] Neuroscientist Joseph LeDoux simply says, "You are your synapses. They are who you are."[4] Regarding our ability to freely control our actions, neurologist Mark Hallett asserts that "the more you scrutinize it, the more you realize you don't have it."[5] As unnerving as these claims may be, the problem goes deeper still.

In claiming that his brain caused him to murder John Collins, Mobley's lawyers were, philosophically speaking, saying that Mobley was not *morally responsible* for killing because he was caused to do so by his brain. We can summarize their argument as follows:

1. Neurobiological interactions in Mobley's brain, facilitated by a deficiency in the enzyme monoamine oxidase A, determined that Mobley kill Collins.
2. Determined actions are not free.
3. One cannot be held morally responsible for actions that are not free.
4. Therefore, Mobley cannot be held morally responsible for killing Collins.[6]

However, the jury did hold Mobley responsible. In the jurors' view, Mobley was faced with a moral choice. After robbing the pizza store, he could have left without firing a shot. If he was worried about being recognized or about Collins calling the police, he could have come in wearing a mask and tied the store manager to a table. However, the jury members decided, Mobley *chose* to kill. They felt justified in blaming him because we hold people legally responsible for their actions, since we assume that people freely choose to act as they do. We assume that every person is a moral agent with the ability to consciously reason through decisions and act according to what he or she believes is right.

But what if the defense counsel was right? What if the complex operations in Mobley's brain caused him to steal? That would mean that his decision was not free, but rather forced upon him by his biological makeup. If everyone's behavior were caused by neurobiological wiring, then no decision could be made freely. It wouldn't matter if you robbed a bank or shot the president, kicked your dog or came late to work—it would all be caused by processes over which you had no control.

If it is the brain that determines our decisions, then each person has at his or her disposal a powerful argument against moral responsibility:

1. Neurobiological interactions in my brain determined that I did X.
2. Determined actions are not free.
3. One cannot be held morally responsible for actions that are not free.
4. Therefore, I cannot be held morally responsible for doing X.

In short, moral responsibility does not exist.

Yet my deepest feelings tell me that every decision I make is mine. I am a moral agent with the power to control my decisions. I have *free will*. It is my conscious faculties that willfully deliberate, often painstakingly, looking at details and considering principles, until I resolve to act, with the understanding that my action will have consequences for which I may have to answer. These feelings are intimately familiar to me and are essential ingredients of my identity.

On the other hand, my knowledge of brain science tells me that within my skull there exists a biological system more intricate than any other. I know that throughout my conscious deliberation, this dynamic system is firing with activity. Billions of networked entities are exchanging and modifying information in ways that can be correlated with my thought and behavior. I know also that I have neither a complete knowledge of this system's operation nor the power to fully control it.

I believe that both my conscious experience and the principles of neurobiology are true, yet when it comes to the making of a conscious decision, they appear to be fundamentally in conflict—and this conflict may threaten our perception of morality and responsibility. One of the main challenges in discussing this dilemma is deciding how to pose the question clearly, and I think that philosopher John Searle has done it correctly. So using Searle's construction, let me put the problem as precisely as I can.

Suppose that at time A, Mobley (or any person we might choose) is faced with a moral choice: to commit murder or walk away. At this moment, assume Mobley is aware of the reasons for deciding one way or the other and is completely mentally competent. Ten seconds later, at time B, he pulls the trigger and kills his victim. Assume that nothing interferes with his mental decision-making process between times *A* and *B*. We are concerned only with Mobley's mind and the choice he makes. If the sum of the neurobiological interactions in Mobley's brain at time *A* is sufficient to determine that, at time *B*, he commits murder, then *free will and moral responsibility do not exist*. He cannot be blamed for his crime because it was determined by his neurobiological circuitry: his brain made him do it. On the other hand, if the sum of the interactions in his brain at time *A* is *insufficient* to determine his decision to commit murder ten seconds later, then it is possible, but not certain, that he had free will and was responsible for what he did.[7]

Our belief in moral responsibility, derived from the assumption that we consciously wield control over our thoughts and actions, seems to form the underpinning of nearly every facet of human life. It shapes the way we treat people around us, the hiring and firing of employees, the academic system of grading, the concepts of crime and punishment, and the relationship between parent and child. It is the origin of pride and guilt, care and fellowship. We can scarcely imagine discarding this idea, one so vital to our understanding of people and society.

Yet, in the shadow of emerging developments in the neurosciences, that essential freedom, or what we think of as a freedom, has come under threat. Neuroscience threatens to demonstrate—some say it has already demonstrated—that our sense of conscious will, that feeling which we know most intimately, is an illusion created by our biological machinery, and that we have no more say in the construction of our future than does a rock in its motion down a hill.

If this claim proves true, affirming the contention of Mobley's lawyers that our choices are determined by the rules of neural processing, how are we to understand moral agency, our moral place in the world? This question strikes at the core of who we are, so to peer at it more carefully is of consequence to every person, if only to better understand ourselves.

2

THE SHADOW OF DETERMINISM

"This is the age of neuroscience—an age of tremendous advances that will exponentially expand our understanding of the brain," says Mriganka Sur, head of the department of Brain and Cognitive Sciences at MIT.[1] Over the past few decades, the field of neuroscience has experienced revolutionary expansion—an explosion of new researchers, inquiries, and findings. As I write this, the membership of the Society for Neuroscience is approaching forty thousand scholars. There are well over three hundred journals that focus solely on the discoveries of neurobiology.[2] More and more lines of research, scientific and otherwise, are beginning to converge on the study of the brain. As one researcher writes:

> The research that is possible, or is already taking place, represents not just an extension of earlier efforts but a qualitative change. From a base of knowledge about the brain in general, neuroscience is now making the first exploratory inroads into the features that characterize us as humans: the ability to create and to calculate, to empathize, to recall and plan.[3]

Neuroscience may be the fastest growing scientific frontier, and advancements are being made every day. Neurological lie detectors are making their way into courtrooms. Prosthetic limbs plug into our brains and are operated by neuronal group activations. Chemical compounds sharpen our minds and alter our personalities. With each advancement, researchers in the field seem to become more certain of their belief that it is the mechanical interactions of the brain, over which we have no control, that dictate our every thought and action.

Why is the rise of neuroscience a threat to human agency? The answer is that the principles of neuroscience as well as the current research emerging from the laboratories provide support for a position known as *neurobiological determinism*, the view that every human thought and action is completely caused, or determined, by conditions in the brain.

For comparison, consider an example of determinism in nature. Suppose that you are standing at the top of a steep, jagged hill. Uneven patches of grass cover its surface. Rocks are strewn everywhere: pebbles, sharp angular stones sticking out of the dirt, and small boulders. The hill is rampant with rotting logs, gnarled tree roots, and clumps of soil. What happens when you toss a stone down the hill? From the moment it falls from your hand, the stone will be part of a series of events that may seem completely random. It starts to roll down straight but then veers in one direction or another. It begins to tumble faster down the hill, bouncing off rocks and exposed roots. It skids down muddy areas, rebounds off of logs, and spins wildly through patches of grass until finally it rolls to a stop somewhere past the foot of the hill.

There appears to be no way you could have predicted the motion of the stone as it tumbled down the hill and landed where it did. The motion of the stone seemed random, but was it? Most people would agree that it was not. The reason, they would say, is that there were an enormous number of variables that caused the stone to move as it did. Those would be things like the size, shape, and weight of the stone; its initial speed; the way it was thrown; the steepness of the hill; the firmness of the soil; the number and types of obstacles that were in its way; and many other factors. From the moment the stone leaves your hand, its every movement—every bump, turn, and roll—is already determined by the conditions in place at that moment. Since you didn't know exactly what those conditions were and how they would affect the motion of the stone, you could not have predicted how it would move or where it would land.

Now imagine that you *did* know all the relevant information about the stone, the hill, the obstacles, and how all these factors would interact to steer the stone down the hill. Since the movement of the stone rolling down a hill is determined by the laws of physics, if you are aware of all the relevant conditions as well as the laws of motion as they pertain to the stone, you should know how it will move and where it will end up. The motion of the stone is *determined*.

This assumption of determinism has been fruitful in science. For physicists, it provides a basis for understanding projectile motion, collisions, elasticity, waves and fluids, electricity, the movements of planets—pretty much everything on the macroscopic particle scale. Chemists have always been able to understand chemical reactions deterministically—mixing baking soda and vinegar under the right conditions will always yield a foaming product. Biologists view all the processes of the body, such as protein synthesis and DNA replication, as determined chains of events.

In our brains, the interacting factors are not rocks and roots, but billions of neurons, neurotransmitters, and other chemicals. Since neurons and their signaling molecules abide by the laws of physics, brain processes can be traced as a determined series of events. Neurons transmit signals to their neighbors by firing an electrical impulse. The firing of the impulse is driven by a voltage difference between the cell and its surroundings. The voltage difference is caused by the existence of an ion gradient across the membrane, such as from a higher concentration of sodium outside the cell than inside. This gradient is created by the work of ion channels that swing open to allow the passage of ions or seal shut to maintain the gradient as is. These channels operate in conjunction with sodium-potassium pumps: energy-consuming mechanical structures that toggle between two conformations in order to ship sodium ions out and haul potassium ions in. The work of the ion channels and pumps is driven by the electrical or chemical signals generated by neighboring neurons—signals that are in turn fueled by voltage differences, ion gradients, and the operation of pumps and channels. Each event in this neurobiological series is caused by prior events.

Considering this chain of causal connections, the neuroscientist says that the brain is a determined system, like almost every other system in nature. It follows, then, that the outputs of the brain are determined by prior processes. Our choices, our desires, our beliefs, our reflections—all are determined by neuronal communication.

If our decisions are determined by the same set of physical laws that govern the motion of the stone, that would mean that each decision we make is the only one we could have made. It would mean that, just as a physicist could predict the motion of the stone, if given enough information about the stone and the hill, a neuroscientist could predict your thoughts and actions—your future—if given enough information about your brain. As the philosopher Pierre Laplace wrote about general determinism:

> Given for one instant an intelligence which could comprehend all the forces by which nature is animated, and the respective situation of the beings who compose it—an intelligence sufficiently vast to submit these data to analysis—it would embrace in the same formula the movements of the greatest bodies of the universe and those of the lightest atom; for it, nothing would be uncertain and the future, as the past, would be present to its eyes.[4]

This idea, often called *Laplace's demon,* is about the implication of universal determinism. If every event in the universe is determined by prior conditions, then it follows that any event can be predicted, assuming one can discover those prior conditions. We are only concerned with neurobiological determinism, but even in that more limited view Laplace's demon is present: with a complete knowledge of your brain activity and all the relevant chemical equations, a neuroscientist, who we might compare to the demon, would be able to calculate what you will think and how you will act.

It should be clear that the concept of determinism is quite different from that of fate or fatalism. We can understand the extent of this difference from a famous old parable in which a Baghdad merchant sent his servant to the market. The servant returned,

> white and trembling, and said, "Master, just now when I was in the marketplace I was jostled by a woman in the crowd and when I turned I saw it was Death that jostled me. She looked at me and made a threatening gesture; now, lend me your horse, and I will ride away from this city and avoid my fate. I will go to Samarra and there Death will not find me." The merchant lent him his horse, and the servant mounted it, and he dug his spurs in its flanks and as fast as the horse could gallop he went.

Curious about the incident, the merchant himself went to the market, saw Death hanging out there, and asked:

"Why did you make a threatening gesture to my servant when you saw him this morning?" "That was not a threatening gesture," [Death said], "it was only a start of surprise. I was astonished to see him in Bagdad, for I had an appointment with him tonight in Samarra."[5]

Whether he runs to Samarra, sails to an island in the Pacific, or digs a big hole and sits in it, the servant is fated to meet Death. There is nothing he or anyone else can do to prevent his fate. If it is fated that you meet the guy or girl of your dreams at a restaurant on Maple Road on April 17, then that is what will happen, no matter what you do or what is going on in the world. If you are driving in the opposite direction, fate will cause you to turn around. If fate dictates that your annoying neighbor Nelson will not die today, then he will live, regardless of how many times you shoot him with your 12-gauge. Fate is the religious or metaphysical idea that no matter what conditions are in place, the fated event will occur.

In contrast, determinism is not a statement about fate; it is a statement about *causality*. To say that human behavior is determined is to claim that it is completely caused by previous events in the brain.

What's more, determinism says that, in a sense, our behavior is the result of a mathematical process. Since human cognition is a determined system, its operation can be understood in terms of strict mathematical principles. Certain causes in the brain necessarily yield certain mental or behavioral effects. Just as two plus three necessarily equals five, an excitation in the frontal lobe together with an activation of the amygdala necessarily yields a certain thought or action. The variables and formulas of brain activity generate our decisions and control our behavior.

The determinists ask us to think of human cognition as a factory. Raw material is sent in, systematically processed, and a product is sent out the other end. The raw material of the mental factory is the set of stimuli received by our sensory receptors: what we see, hear, smell, taste, and feel, as well as the memories and knowledge we are storing. This disorganized array of memories and sensory material proceeds along the assembly line of neuronal processing until a product, some thought or behavior, is manufactured.

The millions of tasks necessary to generate a behavior are spread throughout the giant processing infrastructure of the brain. The results of lower-level processes are sent to be interpreted by higher-level processes. Each level operates according to a strict set of rules. Just as the machinery of a factory has rules built into it—each piece of equipment works in a certain way—the brain has a strict set of operational procedures. It is a processing unit built out of neurons rather than electronics. Just as the heart is a system for effectively distributing blood to the body and the liver is a system for producing bile and regulating biochemical reactions, the brain is a system of generating behavior. Like every other organ in the body, says the determinist, it is a factory designed for producing what human beings need to get around in the world.

Take the case of someone who jumps into a raging river to rescue a stranger from drowning. If the determinist is right, that action is generated by the systematic processing in the mental factory. Stimuli are sent in, such as screams from the river and the perception of a figure in the water, activating the somatosensory cortex, occipital lobe, and auditory cortex. Processing is then initiated in the hippocampus and medial temporal lobe. Memories are unlocked—rapids, danger, drowning, death, rescue. Emotions set in—the fear of personal harm, of failure; the sense of duty. Hormones are manufactured and released throughout the body. Beneath the cortex, the amygdala and the anterior cingulate gyrus light up as their processing begins. All these factors are incorporated into the decision-making algorithm. Should the man jump into the river? Gears are spinning in the frontal lobe. Within moments of perceiving the problem, his brain ignites with activity, making millions of tiny calculations, and generates the moral answer. The rescuer jumps into the river.

As far as neuroscientists are concerned, this is the most reasonable way to think of the relation between brain and behavior. It certainly seems to be the explanation most consistent with scientific research. However, neurobiological determinism seems, at least on the surface, to carry with it some troubling implications for moral agency. If the interactions in my brain, over which I have neither knowledge nor control, determine what I will decide, then any decision that I make is not free. If my decisions are not free and are simply generated by the running of an algorithm, then I cannot be held responsible for their consequences. My deliberations, reflections, emotions, and decisions are at the mercy of deterministic rules. I am just the stone rolling down the hill.

Determinism was the most commonly held scientific view of all natural phenomena until the acceptance of quantum mechanics, a branch of physics that emerged in 1920s.[6] It was during these years that Niels Bohr and Werner Heisenberg went to work in Copenhagen on understanding the movements of electrons in an atom. What came out of their research sent a shock wave through the scientific world. Heisenberg, under Bohr, discovered that one could never know the exact position and momentum of an electron at the same time. The old atomic model represented electrons as circling the nucleus in definite orbits, implying that we could know the location of an electron at any moment. This view of the atom was shattered by the "Heisenberg uncertainty principle" (also called the *Heisenberg indeterminacy principle*), which showed that we cannot ever know exactly where electrons are—only where they are likely to be—even with an infinitely precise measuring instrument. The reason for this is that at this incredibly minute level, the quantum particle level, interactions between particles are not determined—they are *random*. For the first time, it was discovered that there are facts about the world that we cannot know. Albert Einstein could not believe this. He assumed there must be some causes or "hidden variables" that determined the motion of these particles, something Heisenberg and Bohr were missing. He said, "I cannot believe that God would choose to play dice with the universe."[7] But since then, the uncertainty principle has been confirmed many times. Quantum indeterminacy is a fact. The major scientific view now is that indeterminism, or "randomness," is the basis of all physical interactions.

What does this mean for our question? That depends on whom you ask. There are many who think it means nothing. They say this because the principles of quantum physics do not imply that everything in the world is random. Although randomness may dominate at the quantum level, it clearly does not do so at the level of human beings, biological organs, or even cellular interactions. According to quantum theory, if in a game of billiards I accurately hit the cue ball forward, there is a slight possibility that the ball will go sideways. It is very, very, very unlikely, but possible. The same goes if I drop a basketball. We expect the ball to bounce up to a height that is lower than its starting point. Again, quantum mechanics says that there is a ridiculously small chance that it will bounce to a height greater than the original.

Clearly, quantum theory does not play a significant role in understanding physical events at the level with which we are concerned. Furthermore, the randomness implied by the theory does not help the case for moral responsibility. Our problem is that we cannot be held morally responsible for our actions if everything we do is completely determined by events in our brains. However, we also cannot be held responsible for our actions if they are caused by *random* events. Even if we assume that a random, indetermined sequence of neuronal interactions caused the rescuer to jump into the river, he still cannot be held responsible. Random events, over which he neither had knowledge nor control, caused him to do it and, therefore, he does not deserve credit. So we might relate determinism, indeterminism (randomness), and responsibility in the following way:

1. If neurobiological determinism is true, everything we do is completely caused by prior biological events, so we cannot be held morally responsible for our actions.
2. If indeterminism is true, our actions are random and we cannot be held morally responsible for them.
3. Either neurobiological determinism or indeterminism is true.
4. Therefore, we cannot be held morally responsible for our actions.[8]

However, because we established that quantum theory does not imply that everything human beings do is random, we don't need to concern ourselves with it just yet. Since quantum mechanics doesn't do much to help explain human behavior and it doesn't change the implication for moral responsibility, we can set it aside for now. We will do this not because the quantum approach is unimportant but because its relevance will not be visible at this point. We will return to it later in the book.

For now, our challenge is to consider what the ramifications of neurobiological determinism are for moral agency. The rise of neuroscience has cast a shadow of doubt on the age-old assumption that we are each in control of our actions. Research on the brain seems to show that we are merely biological automatons, decision-generating factories, forced to think and act as we do by our neurobiological engines. Does neuroscience show that we are not morally responsible for our actions? Since our understanding of moral responsibility is drawn from the principle that we act according to decisions that we freely make, it seems that the next step in our investigation is to delve deeper into the nature of that supposed freedom.

3

THE ESSENTIAL FREEDOM

In preparation for the impending war between the gods and the giants, the god Zeus decided that he would need a mortal champion to bring him victory. He therefore traveled to Thebes where he disguised himself as the husband of the Theban queen. With her, Zeus fathered a son who would become the greatest of the heroes, Heracles.[1]

The goddess Hera, wife of Zeus, was enraged by her husband's affair with the mortal woman and resolved to torment his new son. In contempt, Hera roused in Heracles a fiery madness that caused him to kill his wife and three sons. For committing this atrocity, Heracles was condemned to undertake twelve onerous labors, among them to slay a nine-headed hydra and capture man-eating horses.

Heracles was punished for the killing of his family, but was he actually responsible? His hands were the ones that did the killing, as any witness would testify. However, we would be inclined to say that Heracles was not responsible for killing his family because his mind was overpowered by Hera's vengeful spell. He did not freely decide to murder, nor did he consciously will himself to carry out that decision. We might say that, since forces beyond his control caused Heracles to murder, he did not act of his own free will. Since he did not act of his own free will, he could not have

35

justly been held responsible. We can conclude, then, that Heracles did not deserve a one-on-one with the nine-headed hydra.

At least in our modern conception, Heracles should not have been held responsible for murdering his family because the act was not freely willed, but caused by Hera's spell.[2] However, even in our time, we have different explanations for *why* we say he did not have free will. Contemporary philosophy provides two contending answers to this question: the view based on *conscious control* and the view based on *alternative choices.*

According to proponents of the first answer, Heracles did not use free will in making the decision to kill his family because that act was not initiated by his conscious mind. His ability to consciously deliberate about whether to act was overridden by forces beyond his control. Under normal circumstances, in which Heracles would have had free will, he would have thought through his decision before acting on it. He would have asked himself: "Do I really want to smite my family? They are such nice people. Sure, they get annoying at times, but is that reason enough to whack them? I don't even think they have life insurance." Whatever his decision would have been, it would have been a decision made freely, as long as Heracles had the opportunity to deliberate in his mind as well as the capacity to consciously initiate his chosen action—without any interference in his thinking.

In this view, free will is *the ability of a person's conscious self to control his or her thoughts and actions.* Let's say that you want to pick up a pen. In your mind you think: "That's it, I can't stand this any longer! I am going to pick up that pen." Your arm reaches out and your fingers grasp the pen and lift it. Free will comes into play twice: the first time when you decide to pick up the pen and the second when you will your hand to move. In both instances, your mind directs what happens. It is your mental volition that marks your intention to act and commands your limb to carry out your decision.

Deciding to pick up a pen is something that tends not to require any deep contemplation, but, under normal circumstances, it is still a free act. It is a free act because you *have the opportunity* to deliberate about it. Nothing prevents you from reflecting on the decision and willing yourself to act accordingly. What's more, it is your conscious self that abstains from deliberation, and that is a choice you make. Whether or not you choose to employ it, your ability to deliberate combined with your mental initiation of behavior constitute free will.

Now imagine that the person reaching for the pen is the president of the United States. He is sitting at a desk, surrounded by politicians and reporters. Picking up that pen will have great significance for him and the country, because it means that he will sign the bill in front of him, agreeing to raise taxes in order to fund a new education plan. But does he want to sign it? He believes that doing so will be for the best, but he is under heavy political pressure not to. He deliberates in his mind. The plan will raise the standard of quality in schools across the country. It is an investment in America's next generation. But what of the rise in taxes? Polls show that most members of his party are infuriated by this plan. So many have called it a waste of money. Will his picking up this pen and signing this document alienate him from his constituency? What about the upcoming election? Perhaps it isn't right to think in those terms. Perhaps he should just listen to his advisors.

The president has free will. His decision is not random, nor is it caused by some external force; it is the result of careful, reflective consideration of the issues involved: his past and future, his relationships, his reputation, and his concern for the well-being of a nation. All of this cautious introspection culminates in the brief movement of his hand.

In the view we have been describing, free will is a mental ability and is present as long as one's mental faculties are intact. Since nothing interferes with the president's ability to contemplate his decision or consciously initiate his action, he has free will. The only condition in which he would lack free will, say proponents of this view, would be one in which some force controls his mind or bypasses it and causes actions without his conscious consent. For example, if a neurosurgeon were to fire electrical impulses at the motor area of the president's brain, innervating his muscles to sign the document, that would not be a free act. Neither would the act be free if an unconscious twitch, such as one from a neurological illness, caused him to sign the bill. The president also could not be held responsible for any signatures made while sleepwalking or in any other unconscious state. Without the ability to consciously control his actions, the president would be no better off than Heracles while under Hera's spell.

None of these situations are very common. People seldom sleepwalk and there aren't too many deranged neurosurgeons running around with mind-controlling electrodes, although admittedly there are a few. However, according to this view, there is one instigator of behavior that could nullify the free will of every human action: the brain. If neurobiological determinism is true, then it is the brain that determines our thoughts and decisions. *The brain controls the mind.* In the current view, if some force controls a person's mind, he cannot have free will. Free will must be wielded by a conscious agent who is not completely subject to some deterministic set of rules or causal forces. Free will is incompatible with neurobiological determinism.

Having the brain determine the mind is like having a spell run the mind. Heracles was not responsible for killing his family while under Hera's spell. He also could not have been held responsible had the act been determined by neuronal events over which he had no control. Similarly, the president could not have signed the bill freely if the decision was caused by his brain. In both cases, interactions in their brains, much like Hera's spell, would have been the force that caused them to act, instead of their conscious minds. Neither man has free will if his behavior is caused by something other than his own conscious initiation.

The view that free will consists of our ability to consciously control our thoughts and actions seems irreconcilable with the neurobiological assertion that neurons determine our behavior. In our dilemma the idea of free will is sandwiched between those of determinism and responsibility. Determinism threatens free will, and nobody can be held responsible for an act that was not done freely. To attack moral agency is to attack free will. To defend moral agency is to defend free will. In the war for or against moral agency, free will represents the battleground.

There is a second, opposing view of free will that holds that it is actually compatible with determinism. This notion, cleverly called *compatibilism*, insists that we still have free will even though all our thoughts and behaviors are caused by our brains.[3] If correct, this view would imply that our dilemma is meaningless. It would mean that neuroscience does not actually threaten free will or moral agency. There is no battle to be fought.

Our account of the will seems fundamentally irreconcilable with determinism, yet the compatibilist sees no conflict at all. How is this possible? The explanation is that the compatibilists have their own definition of free will. They claim that free will is not a conscious ability—it isn't an ability at all. It has nothing to do with the control of thought or behavior, but has to do with the availability of choices.

According to the compatibilist, having free will simply means having alternative choices. It is *the ability of a person to have acted differently should he have had other desires or beliefs.* Let's say that you are choosing between chocolate and vanilla ice cream and you choose chocolate (as you should). The compatibilist says that you freely chose the chocolate ice cream not because your mind controlled the action but because you had the options of choosing vanilla or choosing neither. Nothing prevented you from taking those alternatives. You could have done otherwise, had you wanted to. You had more than one option.

Now imagine that the person offering the choice of ice cream puts a gun to your head and says "Choose the chocolate ice cream or die." In this scenario, says the compatibilist, taking the chocolate ice cream would not be a free choice since you were coerced into making it. Having a gun to your head effectively eliminates any alternative choices, and thus your free will.

Similarly, the president's choice of whether to sign the education bill is a free choice because there are multiple options available: to sign or not to sign. However, if some deranged schoolteacher had threatened to kill him if he didn't support the bill, the president's choice would not have been free.

In the case of Heracles, however, the compatibilist would agree that he does not freely will himself to kill his family. We said that the reason is because Hera interfered with his conscious deliberation and caused him to murder. The compatibilist says the reason is that Heracles had no other options—he could not have done otherwise had he wanted to.

Compatibilists claim that free will is absent any time you could not have done anything other than what you did, whether that be because of threat, physical barriers, or a simple lack of options. To precisely pick out how this view differs from the one based on conscious control, let's consider two more examples.

Imagine that I lead you into a room and seat you on a chair. I tell you that someone will be coming in to speak to you and that you have the choice to meet with him or to leave the room. What I don't tell you is that the chair is covered in super-adhesive glue, so that if you try to get up, you will be forced to stay seated. When the guest arrives, you see that he is an old friend of yours and you are content to stay seated and speak with him. You don't try to get up because you prefer to stay and talk.[4]

In this illustration you consciously decide to stay seated. However, if you had wanted to get up, you would not have been able to. Do you use free will in deciding to sit? The compatibilist says that you do not. Though you consciously decide to sit, this decision is not made freely because you could not have done otherwise if your desires had been different (if you had *not* wanted to speak with your friend). In our view, you certainly *do* use your free will in deciding to stay seated. Your mind freely wills your body to sit and have a conversation with your friend, and that is what happens. It is irrelevant to the nature of your will that there happens to be glue on your seat at the time. True, you have no other option, but that is a problem of limited choices, not limited will.

One final example: Black is a nefarious neurosurgeon who has been secretly hired by the Republican Party to make sure that the Republican candidate wins the upcoming election. Black suspects that one of his patients, Jones, has Democrat tendencies. While operating on Jones's brain, Black inserts a device that monitors political thoughts. The device is wirelessly connected to Black's computer and will send the surgeon any of Jones's thoughts that relate to his voting intentions. It also allows Black to disrupt any neural activity in Jones's brain and initiate a different sequence of neural activities instead. On election day, Jones steps into a voting booth. Black, who is monitoring Jones's brain, decides that if Jones chooses to vote Republican, he will do nothing. But if Jones should choose to vote for a Democrat, Black will use his computer to initiate a neural firing sequence that will force Jones to vote Republican.

Jones chooses to vote Republican, and Black does nothing.[5]

This case is similar to the previous one: a person consciously, and without interference, makes a decision to act, unaware that there are no alternative choices available. Only one action can result: Jones will vote Republican. That's all the compatibilist needs to know to assert that Jones does not freely choose to vote that way. On the other hand, those who believe that free will is a conscious ability contend that, since Black does not interfere with Jones's mental processes at all and Jones is not sleepwalking, brain-damaged, or electrically stimulated while making the decision, Jones *does* act of his own free will.

Of course, if Jones had chosen to vote Democrat and Black had interfered with his mental functioning to cause him to vote Republican instead, then we would agree that Jones did not freely choose to vote that way. However, since he voted Republican and nothing interfered with his mental control over his thoughts and actions, his free will was intact.

The compatibilist view of free will works with determinism because it says that a determined action is free as long as there are no barriers that prevent the agent from doing something else. Choosing the chocolate ice cream might have been determined by your neurobiology, but as long as no one coerced you to take it or locked the vanilla in box and swallowed the key, your choice was free.

Is compatibilism a solution to our dilemma? I don't think so. When I consider the problem of free will, I want to know whether it is my conscious wishes that control my behavior or whether my behavior is determined by the brain just as the movement of a stone down a hill is determined by its environment. Compatibilism merely says that free will means that we have alternative choices when we act. It is true that this is one way to interpret the word *free*, but it doesn't answer the question. It just provides another definition.

I'm not interested in the circumstances in my environment as I decide. If someone puts a gun to my head and coerces me into stealing something, I might argue that I should not be held responsible for the act, but I certainly stole freely. Free will is the extent to which my mind controls my actions, regardless of what choices happen to be available.

To accept compatibilism is to deny that there is a problem of free will at all. Free will just means that you have alternative choices when deciding. There is no conflict between this idea and determinism. To be sure, every act is completely caused by prior conditions, but as long as you could have acted differently, given another set of prior conditions, you acted freely. So compatibilists do not contrast the word *free* with *caused* or *determined* but with *under duress* and *without alternatives*.[6] This view denies that there is any significance to that feeling of intention or *willing* that we have when we act. I have always seen the view not as a solution to the free will problem but as a way to escape it. As philosopher John Searle writes:

When people march in the street carrying signs demanding "Freedom Now," they are not usually thinking about the nature of causation; they just want the government to leave them alone, or some such. And that is, no doubt, an important use of the concept of freedom, but it is not the concept that is central to the problem of free will, at least not as I am construing the problem. Here is the problem: Are all of our decisions and actions preceded by causally sufficient conditions, conditions sufficient to determine that those decisions and actions will occur? Is the sequence of human and animal rational behavior determined in the way that the pen falling to the table is determined in its movement by the force of gravity and other forces acting upon it? That question is not answered by compatibilism.[7]

So many scientists choose to accept compatibilism because it is an easy way to believe in the complete causality of science and still have some concept of human freedom. If you, too, believe in compatibilism, I am almost envious of you. You can be free of the weight of this conflict and need not read any further. For the rest of us, there is a burning problem to confront.

Neurobiological determinism seems to show that our belief in free will and moral responsibility is mistaken. Just as the spell of Hera caused Heracles to act, the neuronal interactions in our brains cause us to think and behave as we do. It would seem that it is not we who are responsible for our actions, whether moral or immoral, but our neurons. It is as if our neurons have us under a spell, just as Hera controlled the mind of Heracles. This is the threat of neuroscience to moral agency.

To most scientists, the arguments for neurobiological determinism are overwhelming, and free will is an illusion. Conscious agency is an antiquated notion that collapses under the weight of the neurosciences, which will show that every aspect of human decision making can be explained by the same deterministic mechanisms that control the motion of a stone down a hill. As Thomas Huxley wrote:

> Consciousness... would appear to be related to the mechanism of [the] body simply as a collateral product of its working, and to be as completely without any power of modifying that working as the steam-whistle which accompanies the work of a locomotive engine is without influence on its machinery. The soul stands related to the body as the bell of a clock to the works, and consciousness answers to the sound which the bell gives out when it is struck.... We are conscious automata.[8]

Faced with this challenge, we must now ask why it is that we believe there is a problem at all. Why do we not just relinquish our ties to free will and responsibility and submit to the projected explanatory power of neurobiological determinism? The answer is that what we, as moral agents, know from having experienced the depths of conscious awareness that there is more to it than the determinist would have us believe.

4

A TEMPEST IN THE BRAIN

Having served nineteen years in prison for stealing food to give to his starving family, Jean Valjean is beginning his new life as a free man. However, marked as a convict by a yellow ticket, he is ostracized by the townspeople and forced to sleep in the streets. Valjean's fortune improves when the bishop welcomes him into his home and provides him with food and a place to sleep. During the night, Valjean steals some of the bishop's silverware and runs away, but the police find him quickly and bring him back to the bishop's house. To Valjean's surprise, the bishop tells the police that he gave Valjean the silverware as a gift. He even rebukes Valjean for forgetting to take the set of candlesticks as well. Once the police leave, the bishop asks Valjean to promise that, in exchange for the candlesticks, he will be an honest man.

Ashamed of himself, Valjean breaks parole and leaves town. He creates a new identity for himself and settles in the town of Montreuil-sur-mer. After inventing a new and more efficient method for manufacturing black beads, the town's major commercial product, Valjean quickly moves to the top ranks of industry. The town is filled with the homeless and impoverished and Valjean does everything he can to help them, including making many charitable donations and providing employment in his factory to people who otherwise would be without work. The extraordinary reputation he develops as a philanthropist, and after saving children from a fire, leads to his appointment as mayor of Montreuil-sur-mer.

It is during this wonderful peak in his life that Valjean (who was known in the town by a different name) receives troubling news. A man by the name of Champmathieu has been arrested, accused of being Jean Valjean. If convicted, this innocent man will go to prison for breaking parole—unless Valjean himself reveals his identity to the world and frees Champmathieu from the false accusation. Plunged into an impossible moral dilemma, Valjean looks inward and begins a maddening process of mental deliberation to resolve what he will do.

His first thought is that all the troubles in his life have been resolved:

> What am I afraid of? Why am I pondering these things? Now I'm safe. It's all over. There was only a single half-open door through which my past could invade my life; that door is now walled up. Forever!... The goal I have been aiming at for so many years, my nightly dream, the object of my prayers to heaven—security—I have gained. It is God's will. I must do nothing contrary to the will of God. And why is it God's will? So I may carry on what I have begun, so I may do good, may one day be a great and encouraging example, so it may be said that there was finally some happy result from this suffering I have undergone and this virtue to which I have returned.[1]

Valjean is comfortable with this conclusion for only a moment before he begins to question himself. How can he go on with his deception? He would be "robbing another man of his existence, his life, his peace, his place in the world; he would be a murderer in a moral sense."[2] How can he remain silent as an innocent man is condemned? This is not that life he promised the bishop he would lead. He gave his word that he would be an honest man. What would the bishop say now? Valjean decides that it is his duty to make this sacrifice, to give himself up to save the innocent man.

Then again, there are other factors to consider. Valjean questions his thinking again. In his reasoning thus far, he neglected a crucial truth:

> I'm sent back to the chain gang; very well, and what then? What happens here? Ah! Here there is a district, a city, a factory, a business, laborers, men, women, old grandfathers, children, poor people! I have created all

this, I keep it all alive; wherever a chimney is smoking, I have put the fuel in the fire and the meat in the pot; I have produced comfort, circulation, credit; before me there was nothing. . . . I take myself away; it all dies. . . . How could I think of turning myself in?[3]

Valjean realizes that he cannot abandon his thousands of workers to unemployment. He can't leave all the families that need his help so desperately. Is he really Jean Valjean anyway? Perhaps not. He has become a new man. And what of Champmathieu? He has stolen before. He stole apples. Sure, he is not Jean Valjean and has never broken parole, but it is not as if his record is clean.

After everything he has gone through in his life—after all the good he has done—how could he allow himself to be sealed in the darkness of a prison cell? After nineteen years of solitude, how could he endure any more? Burning with the agony of the decision, Valjean begins to destroy everything in his house that reminds him of his former self. He throws it all into the fireplace, until he comes to the bishop's candlesticks. Looking at them intently, Valjean reflects on his thinking once again. Long ago, he promised that he would be an honest man. Will he honor that promise now?[4]

In Victor Hugo's classic novel *Les Misérables*, in a chapter titled "A Tempest in the Brain," Valjean agonizes over the decision of whether to turn himself in. He considers all the people the decision might affect, realizing, of course, that there are consequences that he cannot account for. Valjean construes the problem as both moral and personal, and as having implications for the townspeople. Thinking back on his life experience, he formulates his options, reflecting on the outcomes he can think of. Valjean is aware that there are many possibilities he will inevitably miss. He knows also that the set of facts he is working with is incomplete. Many details of his situation are unknown to him. He is aware of traditional moral rules, but is overwhelmed by the difficult question of how to apply them to this case. Of course promises must be kept, such as his promise to the bishop to be an honest man, but what is the honest choice here? That is unclear. Champmathieu doesn't deserve to go to jail for breaking parole, but does Valjean? Breaking parole might usually be wrong, unless parole itself is unfair. When on parole, Valjean wasn't given a fair chance to get back on his feet; he was shunned and mistreated. Does he deserve punishment for running from that? Perhaps the system is at fault. Deceiving the police is wrong in general, but does that rule apply in this case? Valjean's presence in the town protects people from poverty, misery, and death.

The issues and concerns Valjean has to take into account are many. What seems like a choice between two actions might actually have hundreds of resolutions. Rather than turning himself in, he could send an anonymous tip to the police. He could secretly drop evidence on the doorstep of the court that the real Jean Valjean is still out there. He could find a way to show that Jean Valjean killed himself and is lying at the bottom of the ocean somewhere. The list goes on.

What we have here is what I will call a *boundless problem*. There are no set rules from which Valjean can clearly deduce the proper action. The problem cannot be expressed mathematically, using equations and probabilities. For one thing, there are too many unknowns, too many scenarios that cannot sensibly be expressed as simple probabilities. What's more, there is no rational way to quantify experiences, intuitions, emotions, or personal values—all of which must be incorporated into the decision.

In the end, after judging his many options, Valjean decides to turn himself in. I didn't mention that until now because the result of his deliberation is not as important as the deliberation itself. An answer to the question could easily have been generated by flipping a coin. A computer with a random-number generator could have given an answer in no time, but of course it would not have engaged in true moral deliberation. As far as we know, the ability to contemplate, to consciously deliberate about the deepest moral problems, is something only a human being can do, at least for now. Valjean seems to conceptualize the problem in a way that only a conscious being can, unless neurobiologists can prove otherwise. Using the authority of his mind, he can navigate the tempest in his brain.

He can wander the world of his inner thoughts, pausing to reflect on those regions of his mind that he finds most interesting. From the entirety of his knowledge and experience he is able to select only the most relevant aspects to incorporate in his decision. He can reorganize his thoughts by moving them from one region to another and by making connections between disparate concepts, such as abstract moral principles and concrete events that he observes. Through his deliberation, Valjean can expand, shrink, or otherwise alter the structure of his mentality. He can look within himself. He can reflect on his own thinking, question his reasoning, and question the questions he asks on his own reasoning. Valjean can engage in *moral introspection*.

The depth of conscious deliberation in which Jean Valjean engages when making his decision suggests to us that he has freedom of the will. If his decision to turn himself in were determined like, say, the calculations of a computer algorithm, it doesn't seem as if he would be able to contemplate the problem as he does. Determinism, like a mathematical algorithm, manipulates variables according to a strict set of rules. Physics, chemistry, neuroscience—they all work that way. Yet it does not seem that Valjean's deliberation follows such a set of rules. How could his self-reflection be rule based? This I will call the *problem of moral introspection.*

Let's try to explain Valjean's moral introspection neurobiologically. Of course, our knowledge of the brain is not yet advanced enough to provide the complete explanation, but let's try to get the rough idea. First, Valjean's sensory receptors allow him to perceive the news of Champmathieu's sentencing. The receptors fire neuronal signals to the brain that represent the information they receive. In moments Valjean realizes that there is a problem.

Neuronal activity surges in the hippocampus, amygdala, and cingulate gyrus,[5] in portions of the hypothalamus (near the base of the brain) and basal ganglia (internal structure in the center of the brain), and throughout the brain region between the eyes known as the *orbitofrontal cortex*—that is to say, Valjean experiences waves of emotion.[6] At the same time, neuronal systems begin to process a response. Neuronal interactions are amplified in Valjean's frontal lobe and associated regions, such as the basal ganglia, anterior cingulate cortex, and inferior parietal cortex (the region above the temporal lobe). These mechanisms generate an output: the decision to turn himself in. Neural activity proceeds to the motor areas of Valjean's brain to initiate movement, so he can start walking to the courthouse.

Let us suppose, however, that this is more or less what happens in Valjean's brain. The determinist claims that the operation of this system is what causes Jean Valjean to deliberate as he does, or at least make him believe he is deliberating. Could it be that mindless neuronal circuits in the brain, which work just like electric circuits in a computer, cause Valjean's conscious introspection in the same way the surface of a hill determines the motion of a stone rolling along it? Can his mentality be expressed as a mathematical equation, a product of a biological factory? There are those who believe that it cannot—that there is no way that a determined system, one governed by rules and algorithms, could ever attain the kind of conscious introspection that Valjean does. There must be some element of free will, of moral agency, at work.

The question, however, is how moral agency, with its special decision-making power, could arise from the computer-like, nonconscious mechanics of the nervous system. There are two very different and unsatisfactory answers that could be given, one mystical and one scientific.

The mystical answer would be to say that human consciousness, which includes agency as one of its aspects, is fundamentally different and separate from the brain. This idea, known as *dualism*, holds that physical things and conscious things are distinct kinds of substances and belong in separate categories. The view is attributed to the philosopher René Descartes, who lived in the seventeenth century. The idea of dualism is based on the observation that the mind seems very different from physical things. Descartes said that he could imagine himself existing without a body. It was clear to him that there was an elemental difference between his arms, legs, and head, and his thoughts, desires, and feelings. He concluded that the mind is not merely a mechanical bodily process like breathing or digestion.

Dualism has evolved as a theory since then and is still supported by a number of philosophers. As for Descartes' version, he and his dualistic beliefs have now become a stomping ground of scientific ridicule. Whenever there is discussion of the science of consciousness, there is a mention of Descartes' silly, mistaken idea.[7] Neuroscientists assert that the separation of mind and body implies that there is some kind of nonphysical, conscious "spirit" or "soul" that controls the bodies of human beings, a concept mockingly referred to by philosopher Gilbert Ryle as the "ghost in the machine."[8]

Even new formulations of the idea are accused of being wildly unrealistic and symptomatic of an ignorance of science. Personally, though I wouldn't be that harsh, I agree that dualism doesn't hold water as a theory. It does, however, use genuine insights about consciousness in an attempt to reconcile our dilemma. If you believe that mind and body are separate, then you can easily hold that free will and moral agency exist. Our decisions can't be determined by our brains if consciousness is a nonphysical entity separate from the brain—physical processes can't determine nonphysical events.

For some people, dualism solves the problem. In my view, however, the growth of the field of neuroscience shows that dualism is not a strong position to take. It is a problematic concept without scientific evidence to support it. In this book, I reject dualism as an answer to the question.

Some have proposed an alternative explanation, one that could be consistent with scientific principles. It is the concept of *emergence*, or *emergent properties*. Emergence refers to the idea that a system can be greater than the sum of its parts. Consider the chemical elements sodium and chloride. If you throw a lump of sodium into a glass of water, it will violently explode.[9] If you drink a glass of liquid chlorine, you will die from its toxicity. Yet sodium chloride, the combination of these two elements, is something we put in our food all the time: table salt. The property of not being explosive or toxic *emerges* from the combination of toxic and explosive parts. Similarly, neither sodium nor chloride is salty, but the combination of the two is. Saltiness is an emergent property of the combination of sodium and chloride.

Emergent properties are found throughout nature. Oil is greasy, though the chemical elements that compose it are not. Sugar is sweet, but its components don't have the slightest hint of sweetness. Some scientists have proposed that, just as greasiness emerges from nongreasy components and sweetness emerges from nonsweet components, consciousness is an emergent property of interacting nonconscious neurons. Though neurons don't have consciousness or agency, when you put a lot of neurons together and have them interact properly, what emerges is more than the sum of its parts.

It has been claimed that the concept of emergent properties allows for the representation of the mind as distinct from the operations of the brain without resorting to dualism. Many neuroscientists who work on aspects of the consciousness problem embrace the emergent theory of consciousness. Some have gone to great lengths to find experimental evidence to support it.[10]

Is consciousness truly an emergent property of the brain? It may be. The theory is certainly better than dualism. However, as it stands, emergentism doesn't help the case for free will and moral responsibility. Why? The reason is simple: an emergent property may be more than the sum of its parts, *but it is still determined by its parts*. Sodium chloride may have completely different properties than its components, sodium and chloride, but that doesn't mean that the saltiness of sodium chloride is undetermined. No matter what, if you combine sodium and chloride, the product will be salty, nontoxic, and won't violently explode when added to water. The properties of sodium chloride are different from, *but still determined by*, the properties of its components.

It's true that greasiness is an emergent property of a certain organization of carbon, hydrogen, oxygen, and sulfur, but it is a property determined by the interaction of those components. Carbon, hydrogen, oxygen, and sulfur are not in themselves sweet, but when bound to one another in the proper ratio and structure, physical laws determine that the property of sweetness emerges.

In the end, it appears that though the concept of emergent properties seems like a neat solution to the problem, a way to save free will, it provides no way to escape from the clutches of determinism. The only way the idea could be applied in the defense of moral agency is if it were suggested that when consciousness emerges from interactions of neurons, it does so with the property of not being determined. If correct, this theory would imply that moral agency can arise from the determined workings of the brain, but that is a claim that would have to be argued for (and I will discuss such arguments).

In the meantime, we are left to contend with our increasingly inexplicable dilemma. For the determinist position, Valjean's apparent conscious introspection poses a problem because it seems to be a process very unlike the determined processes of the brain and is so far unexplainable neurobiologically, though determinists contend that it eventually will be. The determinist is forced to deny that this introspection, guided by a moral agent, is what controls our actions. For the moral agency position, the problem is figuring out how free will is possible given its obvious dependence on the brain.

Let's start with the determinists. It's easy to say that all our behaviors are completely determined by brain events, but can that be demonstrated? The determinists certainly have a lot of explaining to do. The apparent power of moral agency seems unparalleled in nature; no system has been known to have the unusual properties that some attribute to free will. The common understanding of Valjean's story is that it was only through his introspective, conscious deliberation, of which boundlessness is a feature, that he was able to grapple with the unspeakable difficulty of his situation. Confronted with a boundless problem, consumed by countless unknowns and immeasurable complexity, with nothing to guide him but his experiences, his intuitions, and his own self-reflection, he was able to exert his mental faculties and draw from his inner wealth of knowledge in such a way as to come to grips with his challenge and achieve the most satisfying resolution for his circumstances. If we are to be convinced that all our thoughts and decisions are actually determined by neuronal interactions, the arguments that come our way have to explain how *all of that* can occur without a trace of free will or agency.

To discover whether these arguments mark the end of moral agency, we will have to see them for ourselves and decide how, if at all, they should affect our self-conception. It is at this point that we will turn our attention to the core of the threat to moral agency: the empirical findings of modern neuroscience, the great advances that have brought our question to bear. We will consider the main neuroscientific claims one at a time and decide how far they go in explaining the problem of moral introspection. For each, we must assess the strength of its evidence, the soundness of its reasoning, and the plausibility of its implications. Each chapter that follows will represent yet another attack on the conviction that we are in control of our actions—that we have the power, like Jean Valjean, to freely wield control over the tempest of our thoughts and decisions. Faced with this bombardment, will the moral agent remain standing in the end?

5

NEUROLOGICAL DISTURBANCE

As you are walking down the street, you spot your old friend Eddie Devere coming toward you. You vaguely remember him from college as the guy who loved reading Shakespeare. He smiles and waves. When he reaches you, he smiles again and offers to shake your hand. You ask him how he's been. Suddenly, his facial expression shifts and he says: "Thou art a reeky, pigeon-livered rascal!"

You're caught off guard.

Devere continues. "Thou art a mammering, common-kissing miscreant! A foul and pestilent congregation of vapors! A dog-hearted, elf-skinned canker blossom! A jarring earth-vexing dewberry!"

Much offended by these epithets, and being without a decent comeback, you walk away from your friend, certain that you will not try to interact with him again. Devere was rude to you. You consider him responsible for his behavior because you assume that he freely chose to act that way. In this case, however, you are wrong.

As Devere approached you on the street, he had every intention of being friendly and pleasant. He had planned to greet you and make polite conversation. As he offered to shake your hand, an unbearable urge to insult you came over him. He tried to fight this urge, but it became stronger and stronger. Finally, despite his efforts, he lost control and the Shake-spearean insults were released. Once he heard the words that he had blurted out, Devere was upset that he offended you. He was also sorry that you left before he had a chance to tell you about the damage in his brain that caused him to be rude.

Because of the damage in his brain, Devere's ability to consciously will his actions has been impaired. We might say that he has "lost his free will," or at least a conspicuous part of it.

Devere cannot stop himself from insulting you because he suffers from a disorder known as *Tourette's syndrome*. Tourette's patients are character-ized as having both motor and phonic (verbal) "tics." The motor tics are involuntary and sometimes shocking movements of the body. These can include things like head and shoulder jerks, facial grimaces, and even self-harm. The phonic tics range from simple utterances made at random to what is termed *coprolalia*. The name is from the Greek words *copra* for "feces" and *lalia* for "speech." Patients with Tourette's may spontaneously yell profanities, insult people (though usually not in a Shakespearean fashion), or make other inappropriate comments. Patients describe their tics as uncontrollable urges that they cannot help but express.

The question, however, is what went wrong in Devere's brain that caused him to behave badly? Tourette's results from problems with the basal ganglia, a group of nuclei beneath the cortex near the front of the brain. Some scientists theorize that the disorder stems from poor commu-nication between the basal ganglia and the frontal lobe.[1] This problem in Devere's brain affected his ability to control his behavior and caused him to insult you.

There are some who have appealed to cases like that of Devere to devise an argument against free will and moral agency. We might state that argument as follows: Since the brain, in a damaged state, determines Devere's insulting behavior, it must be that the brain, in a healthy state, determines his normal behavior. Thus, the entirety of Devere's behavior is determined by whatever state his brain is in. Devere's behavior is not a matter of will or agency or deliberation. It depends only on the interaction between regions of his brain such as the basal ganglia and frontal lobe. To assess the potency of this argument, we must examine the evidence from which it gathers its momentum: clinical cases.

There are many examples in medicine of the correlation between damage to a patient's brain and reported dysfunction of his conscious will. Take, for instance, Parkinson's disease, infamously known for its character-istic physical tremors. One will typically see a Parkinson's patient having hands that shake uncontrollably, movements that are certainly not initiated by the patient. What is not well known about Parkinson's is that it some-times involves problems in cognitive function: difficulties in directing attention, evaluating situations, interpeting social cues, and controlling impulses have been known to accompany the Parkinsonian tremors.

Problems with willful control of the body are much more apparent in Huntington's disease, also called *Huntington's chorea*. The word *chorea* (Greek word for "dance") refers to the jerking, uncontrollable movements that are symptoms of the illness.

Another, even more dramatic, example is *alien hand syndrome*, in which a patient's hand might spontaneously grab a nearby object without the person willing it to do so. Sometimes the patient is unable to make the hand release the object and may have to use the other hand to remove it. One patient found that he could make his "alien" limb let go by screaming at it.[2] Another patient said that her alien hand tried to strangle her.[3]

So far, we have seen how damage to the brain can result in a person exhibiting involuntary movements, but this is not the only connection we can draw between the unhealthy brain and the unhealthy will. Brain injury can manifest itself in an enormous spectrum of symptoms that appear to be disorders of the will. The ones we will consider include problems with initiating action, inhibiting action, and the experience of will. It is through providing this multitude of examples that the determinist can strengthen his case that the correlation between the dysfunctional brain and dysfunc-tional behavior shows that free will does not exist.

It is often said that Hamlet's tragic flaw was "weakness of the will," a supposed aspect of his personality. There are, however, actual patients whom we might characterize, due to psychiatric disorder, as having a weakness of will. A tragic example is a syndrome known as *akinetic mutism*, in which patients are unmoving (akinetic) and unspeaking (mute). They are unable to communicate or carry out purposeful behavior even though they sit with their eyes open and seem to be aware of what is happening around them.

In one fifty-seven-year-old patient, testing showed that, in terms of his physical capacity, he should have been able to speak and move normally.[4]

Nothing was wrong with his muscular control: he would instantly withdraw his hand from a painful stimulus. He blinked when objects moved quickly past his eyes and he was able to visually track the movements of people and objects in his environment. The patient was clearly aware of events in his surroundings, but when doctors tried to interview him, he did not respond to any questions or commands, though he would look directly at them. He would not leave his bed and would not feed himself.

This illness, which results from damage to parts of the frontal cortex, demonstrates the causal relationship that exists between impairment of that brain region and a severe dysfunction of willful control: the ability to initiate actions. Human agency, however, allows us not only the capacity to initiate, but also to *inhibit* our behavior. Well aware of this, the determinist points us to clinical cases that suggest this aspect, too, might be generated by the mechanisms of neurobiology.

Vladimir was a young engineering student in Moscow when he was hit by a train after stepping on the tracks to retrieve his soccer ball. As a result of the accident, he experienced severe damage to his frontal lobes and, thus, to his executive functioning. Vladimir resembled other patients with frontal damage in having trouble following instructions, but he also developed another, more interesting symptom: he had trouble *stopping* to follow those instructions once he began. When asked to draw a circle on a sheet of paper, he would sit idly until the examiner grasped his hand and nudged him to begin. He would draw the circle, but then he would draw another, and another. He would continue drawing circles until the examiner took his hand and pulled it away from the page.

Another demonstration of this inability to inhibit his behaviors came when Vladimir was asked to repeat this simple version of the story "The Lion and the Mouse":

> A lion was asleep and a mouse was running around him making noises. The lion woke up, caught the mouse, and was about to eat him up, but then decided to show mercy and let the mouse go. A few days later, hunters caught the lion and tied him to a tree with ropes. The mouse learned about it, ran down, gnawed the ropes and set the lion free.

When instructed to retell the story, Vladimir said:

So the lion made friends with the mouse. The mouse was caught by the lion. He wanted to strangle him but then let him go. The mouse started dancing around him, singing songs, and was released. After that the mouse was accepted in his house by ... lions, various animals. After that he was released, so to speak, he hadn't been captured, he was still free. But after that he was completely released and was walking free....

At this point the examiner cut in to ask whether Vladimir was finished. He replied "Not yet" and continued to ramble on:

So, he was released by the lion completely, after the lion listened to him, and he was released to go to all the four directions. He didn't run away and remained to live in his cave. Then the lion caught him again, some time later.... I don't remember it exactly. So he caught him and released him again. Now the mouse got out of there and went to his hangout, to his pad. The mouse goes on and on and talks about his pad. And there is another mouse there. So the mouse opens the door to this... what do you call it? Hi! Hi! How are you doing! OK, more or less. I am all set. Glad to see you. I have an apartment... and a house... and a room. The bigger mouse asks the smaller one: How are you doing? How is it coming along?[5]

Eventually, the examiner left the room—not that I can blame him. Vladimir has a mental disorder resulting from damage to his frontal cortex, and this has caused him to lose the ability to inhibit his behavior. Perhaps his successful inhibition of behavior before the accident was determined by the proper working of his frontal cortex.

A more familiar example of this kind of syndrome is *obsessive-compulsive disorder* (OCD), in which patients are unable to prevent themselves from engaging in certain compulsions even if they know that what they're doing is irrational. They might, for example, wash their hands hundreds of times a day or feel compelled to have everything around them in a perfect arrangement. Whatever the compulsion, people with OCD may be unable to stop themselves from giving in to it. In this way they share Vladimir's difficulty of using the will to terminate a behavior. They also share another characteristic with Vladimir: problems with the frontal lobe.[6]

The compulsions associated with OCD are distinctive in that they are repeated constantly regardless of the person's environment. What I mean is that a person who compulsively washes her hands will feel the need to do so not only while she is near a sink or a bar of soap but all the time. The act of washing her hands is not necessarily a response to an outside stimulus, in contrast to the compulsions that we will address next.

In a Paris hospital, Dr. François Lhermitte became known for demonstrating how patients with frontal lobe damage can be controlled by objects or events in their surroundings.[7] This phenomenon is sometimes termed *environmental-dependency syndrome* and results from damage to the executive function of inhibiting oneself from acting. In one experiment, Dr. Lhermitte took two frontal lobe patients, a woman and a man, one at a time to a bedroom containing an unmade bed. Upon being let into the room, the woman immediately hurried over to make the bed, though she had never seen the bed or the room before.

The man was led into the room next. When he saw the bed, which was unfamiliar to him, too, he ran over to take a nap in it (typical male!). I admit that I often feel the urge to hop in a bed and take a nap, but I am able to control that urge, especially when the bed belongs to someone else. It appears that, as a result of frontal lobe damage, these two patients did not have the ability to exert self-control in that way. They knew how to use the bed but could not properly evaluate their situation (an important component of executive function) and act according to what was appropriate at the time.

Dr. Lhermitte also tells of a different kind of meeting with one of his frontal lobe patients. On a table near the entrance to the room, he set up a framed picture with a hammer and a nail. When the patient arrived and noticed the objects, he banged the nail into the wall and hung the picture on it, though he was, of course, not asked to do that. In another case, which I find somewhat disturbing, Lhermitte placed a hypodermic needle in front of a patient, dropped his pants, and turned to reveal his backside. The patient's executive dysfunction caused him to miss the appropriate course of action—to ask the doctor why he took his pants off or run away and call some kind of medical abuse hotline—and to instead grab the needle and stick his doctor in the rear. In this case, however, the experimenter's mental health may be as questionable as his patient's.

The utilization behavior of Lhermitte's patients is one kind of environmental-dependency syndrome. Another is called *forced hyperphasia*, a disorder involving compulsory, involuntary speaking. A demonstration of this disorder came from an eighty-four-year-old woman who initially only had a problem with uncontrolled imitation.[8] When her doctors would wave, she would wave. When one of them touched his chin, she would do the same, even after being asked to stop. When asked why she imitated everyone, she replied that she felt she had to.

The imitation behavior was gone after two weeks and replaced by forced hyperphasia. Now, she constantly named objects and described events in her room. When a doctor touched his chin, she would, rather than imitate him, yell something like "The doctor is touching his chin!" Note that the patient didn't repeat what her caretakers said; she just continuously described whatever she saw around her. When asked to explain her behavior, she again replied that she felt she had to.

So far we have discussed the different ways in which willful control of the body can be disturbed, but there is one aspect of conscious will we left out: the way it feels to control our behaviors. We would all agree that willful decisions are associated with a certain feeling or experience. We sense that we mentally command our limbs to move as they do. The feeling, for example, of willfully bending one's knee is very different from that of a knee reflex that a doctor might induce with a rubber hammer. We feel like we are somewhere in our heads, making decisions and controlling our behaviors—at least that is how it feels to most people. Some schizophrenic patients, however, have a very different experience of will, a symptom that neuroscientists have traced to disturbances in brain biology—yet another affirmation, according to the determinist, that every aspect of this "free will" can be explained fully as an algorithmic, deterministic process in the brain.

Schizophrenia is a disease that often causes strange hallucinations. One such symptom is what we might call *I-disorder*.[9] Patients who suffer from it may feel that their thoughts no longer belong to them. They feel as if their minds are elsewhere, far from their bodies, or that their ideas are being inserted into their heads by someone else. They describe the feelings by saying things like "I feel guided by a female spirit who has entered me" or "An electric remote control is controlling me."

Other forms of the disorder cause patients to feel that their thoughts can control things that they in fact cannot control. They might claim to be able to control the movements of cars (without driving them, that is) or that of the sun. A brain-imaging technique called *positron emission tomography* (PET)[10] was used to study patients with this kind of hallucination.[11]

The results of this study showed activation in the anterior cingulate cortex (between the front and top of the brain, just below the surface) and the right inferior parietal cortex (upper right side, just above the temporal lobe) while the patients engaged in willed movements. Not surprisingly, the strange experiences of will described by these patients could be traced to a neurobiological source.

In the case of your friend Devere, who has Tourette's syndrome, a scientist might conclude that had Devere been free of the disorder, his normal behavior would have been determined by his healthy brain. He probably would have asked about your life and your family and your interests, but that would have been just as determined by his healthy brain as his insults were determined by his disordered brain. Either way, his behavior was beyond his control. Either way, he could not have been held morally responsible for his actions.[12]

All the patients we have discussed exhibit an impairment of conscious control that can be correlated with damage to a certain part of the brain. They, along with Devere, serve for the determinist as a vast body of evidence that human beings are not moral agents, with the power to think and act as they choose. The words *free will*, if they should be used at all, simply refer to a set of brain processes—electrical impulses, neurotransmitter releases, signal transduction—not to the conscious control of decision making. Since disabling parts of the brain disables free will, the determinist concludes that what appears to be free will must actually be just the operation of those parts. Thus human decisions cannot be free; they are caused by the deterministic algorithms of the brain.

Through further research, the determinist contends, we will uncover the precise mechanism by which our neurons cause our behavior. We will be able to map out the process in all its complexity. Human decisions will become predictable. Mathematical equations will model our cognitive processes. Lurking somewhere in the brain is the secret of conscious will. It is waiting to be found.

6

THE SEAT OF THE WILL

To scientists of the seventeenth century, the appearance of fire and rust were mysterious, unexplained phenomena. They wondered, for example, what property in a block of wood allowed it to catch fire. They grappled with the problems of why flames burned for a certain amount of time before dissipating, why burning wood would leave behind ash, and why metal had the capacity to exude the reddish brown residue known as rust.

In 1667, after years of scientific bewilderment, a theory arose that promised to fill in the gaps. The theory, originally proposed by Johann Joachim Becher, suggested that the capacity for an object to catch fire or rust is derived from a certain substance contained within it known as *phlogiston* (from the Ancient Greek word *phlogios*, meaning "fiery"). Phlogiston, it was believed, has no color or smell or taste, making it practically impossible to detect. A level of phlogiston is present in all flammable materials—this is what allows them to catch fire. When an object, say, a slab of wood, burns, phlogiston is released into the air. As the burning continues, more and more phlogiston is used up until, finally, when the reservoir of phlogiston in the wood is depleted, the flames die out. Similarly, the loss of phlogiston from a metal rod causes it to rust. Both the rusted metal and the burnt wood, which has been degraded to ash, were thought to become "dephlogisticated."[1]

Nearly a hundred years went by before it was realized that the phlogiston theory was false. The true explanation, which was discovered later, is that the burning and rusting of materials are explained by oxidation reactions in chemistry, those that involve the loss of electrons from a compound.[2] What was for years assumed to be the result of some invisible, undetectable substance turned out to be a very natural chemical process. Looking back, scientists see the phlogiston theory as a primitive theoretical invention symptomatic of the time's limited understanding of chemistry and physics.

Many neuroscientists today make the same accusation against moral agency. They contend that, like phlogiston, conscious agency is the invisible, undetectable force that people concocted as a theory ages ago, long before the rise of neuroscience. It is invisible because it refers neither to physical particles nor to chemical compounds. It is undetectable because, as far as we can tell, there is no way to identify its presence except to ask someone: "Do you feel that you are a moral agent?" and wait for his response—but this is not the scientific way of demonstrating truths.

If someone were to study the process of food digestion in a human being, there would be a few straightforward ways to do so. One could, for example, do a dissection on a cadaver and look at the structural components of digestion, such as the stomach and gall bladder and esophagus. One could investigate how all the parts work together, how the muscular tissue in the esophagus allows it to carry out peristalsis to push food toward the stomach or how the chemical composition of bile facilitates the breakdown of fats. The same type of investigation might be applied to a study of the circulatory system. Looking at the muscular structure of the heart and the network of blood vessels can tell us whatever we need to know about how blood flows throughout the body.

This approach, however, does not seem to work for agency. Nothing in the structure of neurons or the cerebral cortices has yet revealed the secrets of the will. The method of scientific research is to collect data based on observed events. We know that bile breaks down fat because scientists have mixed bile and fatty acids together in test tubes and observed the result. In the case of moral agency, however, there is nothing to observe. It appears that we are no more justified in claiming that a person has free will or agency than we are in claiming that a plank of wood contains phlogiston.

The solution proposed by neurobiologists is to dispose of this ancient doctrine of decision making and replace it with the neuronal model. Through further research, scientists claim, we will discover the seat of the will in the brain. When that day comes, we will be forced to conclude that what we call "freedom of the will" or "moral agency" is, contrary to whatever we may believe about ourselves, just the deterministic interaction of neurons and related chemicals in the cited region of the brain. Judging from the state of neuroscientific research today, that day might arrive sooner than we expect.

The first step in locating the seat of the will has already been taken: analyzing cases of brain damage in which the will seems to be affected. Studies of patients with lesions in specific brain areas allow scientists to correlate the damage with behavioral symptoms the patients exhibit. For example, if we have thirty patients who have trouble perceiving three-dimensional objects and, after using scanning tools, find that most have damage to their occipital lobes (a region in the back of the brain), then we might conclude that there is some correlation between the functioning of that lobe and visual functioning. The same approach has been taken to investigate the mechanics of decision making.

Tourette's, Parkinson's, and Huntington's result from damage to the basal ganglia, a group of nucleii associated with the frontal lobe. Alien hand syndrome results from damage to the anterior cingulate cortex, part of the frontal lobe. Akinetic mutism is caused by trauma to parts of the frontal cortex. The symptoms that afflicted Vladimir were tracked to injuries he sustained in his frontal lobe. Scientists have linked environmental-dependency syndrome and OCD to disorders of the frontal lobe. Studies of schizophrenia have traced its strange symptoms to the dysfunction of several brain areas, the frontal lobe being one of them.

Is the seat of the will in the frontal lobe? It's possible—even likely—but scientists must, of course, delve more deeply into the workings of that system in order to confirm this choice. For now, neuroscientists tend to refer to the frontal lobe as the center of executive function, which refers to the brain processes that, scientists say, coordinate actions that reflect the knowledge and goals stored in memory. It also functions to monitor and evaluate the information it receives from the rest of the brain. It is the top level of brain processing—the computer program that interacts with all the lower-level programs.

The predominant view of neuroscientists is that we have executive functioning but not free will. What we call "free will" is actually the deterministic information processing of the frontal lobe. Many neuroscientists argue for this simply by pointing out the likelihood that the seat of the free will is in the frontal lobe. Since what we think of as human agency depends on the operation of the frontal lobe, they claim, *agency must be equivalent to the operation of the frontal lobe.* And the operation of the frontal lobe is certainly deterministic, a program of executive functioning that endows us with no more free will than a circuit provides to a lamp, an organ stop to an organ, or a jagged hill to a falling stone. The seat of the will implies the end of the will.

As we sit and consider what will become of free will and moral responsibility once the seat of the will is discovered, or even now, the question we must ask is whether the argument of the neuroscientist is sound. If we agree that, one day, scientists will find the system in the brain that allows for conscious decision making, does that imply that what we call "free will" is only the mechanical operation of that system?

The answer is no. There is a logical fallacy here, and it might be more visible if we state the neuroscientist's argument more concisely:

1. What we call "free will" is made possible by the operation of the frontal lobe.
2. Therefore, what we call "free will" is identical with the operation of the frontal lobe.
3. The operation of the frontal lobe is not free.
4. Therefore, what we call "free will" is not free.

The preliminary conclusion in step 2 does not follow from the premise in step 1. Just the fact that there is a neurological correlate of free will (whether the frontal lobe, or something else) does not imply that free will *is the same* as the neurological correlate. We all agree that free will, if it exists, is made possible by the brain. There is no doubt that, without a healthy brain, we would not have free will. That doesn't mean that free will is equivalent to the operation of the frontal lobe—only that it depends on it.

Logical fallacies like the one in the above argument are surprisingly common in the attempts of some scientists to disprove the existence of free will and agency. In fact, we've seen them before. Consider the one in this argument:

1. Damaged brains determine abnormal behaviors.
2. Therefore, undamaged brains determine normal human behavior.
3. Human beings have either damaged or undamaged brains.
4. Therefore, human behavior is determined.

Premise 1 is certainly true. We have encountered dozens of examples demonstrating that an impaired brain can determine impaired behavior. However, it does not follow from this that the healthy brain determines healthy behavior, as step 2 of the argument would lead us to believe. That is like saying "Since a damaged engine in a car determines that the driver will parallel park dreadfully, a working engine must determine that he parallel parks wonderfully." From personal experience, I can assure you that this is not the case.

Again we see that simply correlating the operation of the brain with human decision making is not enough to prove that we do not utilize free will when we act. That only shows that our thoughts and behaviors *depend* on the operation of the brain in some way—but there was never any doubt of that. However, the reliance of our conscious will on the frontal lobe does suggest that our brains have at least some *influence* on our conscious decision-making process. But what is the extent of that influence? Are we truly moral agents with the power to control our behavior, or are we being controlled by our brains?

7

THE SOMATIC-MARKER
HYPOTHESIS

One summer morning in Vermont, a construction foreman named Phineas Gage begins his day of laying track for the Rutland & Burlington Railroad. Specifically, Gage's job is to detonate the rocky terrain of the area to allow the rails to be placed on flat ground. To set a charge in a rock, Gage first drills a hole in it and fills it with explosive powder. Next he inserts the fuse and covers the hole with sand to contain the explosion. He tamps down the sand in the hole with an iron rod to make sure the hole is sealed and the explosion is not directed away from the rock.

Late in the afternoon, Phineas Gage begins to set a charge. He drills the hole and adds the powder and fuse. He tells one of his workers to cover the hole with sand. Someone calls him from behind. He turns briefly. Before his worker has covered the hole, the distracted Gage pushes the iron tamping rod into it, directly onto the fuse and powder. The charge goes off violently. The iron rod shoots upward, passing through Gage's head and landing more than a hundred feet away.[1]

The construction crew runs over to find Gage on the ground with blood leaking from two big holes, one in his left cheek and the other at the top of his head. Shockingly, however, Gage is still alive! After a few brief convulsions, he sits up and begins speaking. Quickly, Gage is taken to a hospital where a physician is able to stop the bleeding and treat the wounds. The doctor is able to prevent infection and, two months later, Phineas Gage is declared cured.

However, soon after Gage is released, it is clear that something is terribly wrong. Before the accident, everyone knew him as being polite, hardworking, and responsible. He had been a model employee and simply a pleasant person to be around. Now, Gage seems viciously angry all the time. His social skills are gone. He is constantly spewing the harshest profanities and insulting everyone around him. Most important, Gage cannot plan ahead. He cannot make sensible judgments about how to act. Instead, he might rattle off long lists of incomprehensible ideas that he quickly forgets. Gage has lost his ability to make good decisions. It is not long before he is fired from his job at the railroad. As his friends quickly recognize, Gage is "no longer Gage."[2]

The above clinical case has become famous in neurology because of the extraordinary mystery it presents: what happened during the accident that so altered the mind of Phineas Gage? In what follows, we will try to answer this question and, in doing so, find powerful implications for moral agency and the nature of the will.

The first thing to figure out is what part of Gage's brain was damaged in the accident. During the explosion, the iron rod passed through his left cheek and came out through the top of his head. In between, the passage of the rod obliterated a part of Gage's brain known as the *orbitofrontal cortex*, located in the lower part of the frontal lobe, just above the eyes.[3] Not much is known about this region of the brain. What we do know is that the orbitofrontal cortex participates in the intricate system of emotional processing that includes areas like the amygdala and the cingulate gyrus. So its primary function can be said to be triggering emotion. But Gage's problem seems to be more than just an emotional one. It's not as if he was simply sadder or angrier or more frustrated than usual. Gage lost his ability to deal with social situations, to come up with reasonable ideas, to judge circumstances appropriately, and to plan for the future. He lost his capacity for rational decision making.

We said that the primary role of the orbitofrontal cortex is to facilitate emotional processing. Yet when this area of Phineas Gage's brain was destroyed, the damage manifested not only in his emotional changes but also in his lack of rational ability. According to scientist Antonio Damasio, we see from this case that there is a powerful connection between the human capacities for emotion and reason.

With regard to decision making, we tend to distinguish between the emotional and rational faculties. A rational decision is one based on *reason*. When emotions get too involved in assessing how we ought to act, it is widely thought that the resulting decision is less valid, less rational. It is thought that emotion is a barrier to good judgment that the rational person must overcome.

Damasio disagrees. He not only rejects the idea that reason must overcome emotion; he says, furthermore, that reason *depends* on emotion. Emotions play an integral part in the making of many, if not all, of our conscious decisions. Initially, this concept may come as no surprise. We make decisions all the time that have emotional components. I might decide to take a vacation from my job because it makes me depressed. I might decide to work twice as hard at my job because I am jealous that my colleagues make more money than I do. In making either of these choices, I am aware that I do so because of my emotional state. Of course emotions are involved in the decision-making process.

However, that is not exactly what Damasio is saying. He is not merely claiming that emotions can play a role in human judgment. That much is clear. Rather, he is saying that *decision making is itself an emotional process*. At the core of this theory, which Damasio calls the "somatic-marker hypothesis," is something not at all intuitive: emotional factors influence our decisions *whether or not we are aware of it*.[4]

Suppose that I am trying to decide whether to confront a friend about his drinking habits. How do I make the decision? As I think about the choice, I consider each possibility and try to foresee the consequences of choosing it. If I were to confront him, he probably would act stubbornly. He would probably resist my suggestions and even resent me for trying to intrude into his chosen lifestyle. Then again, my advice might have an impact on him and prevent him from some difficulties in the future. If I decide not to speak to him, his problem could get worse. He might end up driving drunk or getting into a bar fight. His alcoholism has already gotten the attention of his supervisor at work and could potentially get him fired from his job if the situation doesn't improve.

These scenarios rapidly flash through my mind, too fast to precisely define the details. Each one strikes me in a certain way. This choice doesn't feel right. The other one just seems better. I have certain impressions or gut feelings about which choices are better or worse, even before I actually try to sort out the pros and cons of each. Where do these scenarios, gut feelings, and impressions come from?

Damasio asserts that they come from *somatic markers*.[5] Every time we have an experience, Damasio says, there will be some feelings or bodily states associated with it.[6] These feelings are impressed upon the nervous system and remain connected to the memories of the events as bodily markers. These are what Damasio calls somatic (*soma* is Greek for "body") markers, biological remnants of emotional states. So, for example, if you were to have a frustrating experience doing office work or a disgusting experience eating boiled turnips, somatic markers might be left in your body that associate office work with frustration and boiled turnips with disgust.

As we consider ideas and possibilities, somatic markers are activated, Damasio says. Before we get a chance to evaluate a choice in depth, the effect of the somatic markers takes place unconsciously, influencing how we choose and even *what choices we are considering in the first place*. Some choices may have already been eliminated before you begin your evaluation of the possibilities. As Damasio puts it:

> What does the *somatic marker* achieve? It forces attention on the negative outcome to which a given action may lead, and functions as an automated alarm signal which says: Beware of danger ahead if you choose the option which leads to this outcome. The signal may lead you to reject, *immediately*, the negative course of action and thus make you choose among other alternatives. The automated signal protects you against future losses, without further ado, and then allows you *to choose from among fewer alternatives*. There is still room for using a cost/benefit analysis and proper deductive competence, but only *after* the automated step drastically reduces the number of options. [Emphasis in the original][7]

Before we even get a chance to start our assessment of each choice, the somatic markers have already done their work—and we might be completely unaware of their influence.

Damasio maintains that dysfunction of the somatic-marker system, brought about by lesions or areas of damage in the orbitofrontal cortex, will severely impair one's ability to make decisions. The orbitofrontal cortex is located in the frontal lobe and thus is involved in the brain's system of executive function.[8] The close proximity of these regions in itself provides a measure of support for the idea that somatic markers influence the decision-making process. More powerful evidence, however, comes from cases of damage to the frontal cortex, such as that of Phineas Gage. But his story is only one example, and we will need to find more of them if we want to substantiate Damasio's hypothesis.

Damasio relates the story of one of his patients, whom he calls Elliot.[9] After surgery to remove a tumor from his frontal lobe, it was clear that there had been a profound change in Elliot's personality.[10] Once a good husband, father, and businessman, he now couldn't be trusted to get anything done. He was unable to get ready for work on his own. He couldn't keep to a schedule or properly make arrangements to complete his assignments. After losing his job, he made a series of ill-advised financial and personal choices that led him through bankruptcy and multiple divorces.

Elliot's inability to make sensible decisions was especially shocking to people because he came across as being perfectly capable and intelligent. His memory was intact and he had a solid grasp of what was going on around him. To figure out exactly what was wrong with Elliot, doctors would have to run tests.

Tissue was removed from Elliot's frontal cortex. The first question to ask was whether Elliot had a classic frontal lobe syndrome, that is, some sort of executive dysfunction. Recall that patients with frontal lobe damage typically have trouble initiating and inhibiting actions, difficulties that surely impede decision making. In fact, Damasio initially thought that this was his problem. To test this diagnosis, Elliot was given a well-known test of frontal lobe dysfunction, the Wisconsin Card Sorting Task.[11] The test involves, as you might have guessed, a deck of cards—but not a standard one. The cards in the deck differ from one another in three ways. Each has a certain type of symbol (circles, squares, triangles, etc.), a certain number of symbols, and a certain color. So, for example, one card might display three red squares while another has two blue triangles and another has four green triangles. No two cards are exactly the same, but they may share characteristics. Thus, there are three ways to sort the cards: by shape, by color, or by number.

To begin the test, the experimenter sets several cards out on the table. The subject is told to draw one card at a time from the deck and match it to one of the cards set out on the table based on a sorting rule. So if the sorting rule is to organize by number, then a correct match would be three red circles and three green triangles. The color and shape would not be relevant in this case. But here's the catch: the examiner doesn't tell the subject what the sorting rule is; the subject has to figure it out through trial and error. After each card placement, the experimenter says whether the match is correct or incorrect.

Guided by the experimenter's feedback, a subject who figures out the sorting rule will continue to make correct matches. But then comes the second, more devious catch: after the subject makes ten correct matches, the experimenter changes the sorting rule *without telling the subject*. Suddenly, the strategy that worked before is invalid and the subject must once again use trial and error to figure out what to do. The subject must be able to both discover the initial sorting method and also be able to adapt to the new ones.

Neurologically healthy people have no trouble succeeding at this task. They quickly realize what they need to do and are able to keep up with the changes in sorting rules. Patients with frontal lobe syndrome, however, have great difficulty with the task. They often display a symptom known as *perseveration*, in which they cannot initiate a change in behavior.[12] Such dysfunction is revealed by the second part of the Wisconsin Card Sorting Task: frontal lobe patients are unable to switch from carrying out the current sorting rule to carrying out the new one. Rather, they perseverate— they continue to follow the current rule and cannot change.

Frontal lobe dysfunction, as detected by the Wisconsin test, definitely can cause problems with decision making, even problems just like Elliot's. However, as it turns out, when Elliot was given the test, he did very well! This deepened the mystery for Damasio and his colleagues. Apparently, Elliot was not afflicted by typical frontal lobe syndrome. But what else could it be? Since the brunt of the damage in Elliot's brain was dealt to the orbitofrontal cortex, Damasio guessed that Elliot's problem must be an emotional one.[13] Perhaps it was even a problem with the somatic-marker system. To demonstrate this, however, a new test would have to be developed.

Interestingly enough, this test would turn out to be another card game, called the *gambling task*, which would be designed to mimic the uncertainty and possibilities of reward and punishment that a decision maker must face in real life.[14] Here's how it works: A subject is given $2,000 of play money (though it looks genuine). He is seated at a table on which four decks of cards, labeled A through D, are placed. The subject can draw one card at a time from any deck. Written on each card is an amount of money that the subject will either gain or lose. One card might say that he gains $50 while another says that he loses $100. The subject is told that his goal is to accumulate as much money as possible by the end of the experiment.

What he isn't told, however, is how the four decks are set up. Decks A and B are stacked so that the cards that add money do so $100 at a time while decks C and D only add $50 at a time. Not surprisingly, most patients and healthy subjects initially show a preference for decks A and B because of the higher gain. However, the sensible subject quickly realizes that C and D are the better decks to choose from. This is because of the differences in the cards that take away money. In decks C and D, those cards tend to deduct under $100 when they are picked. But in decks A and B they take away much more, sometimes up to $1,250. The potential for loss in decks A and B far outweighs the potential for gain. The prudent tactic, therefore, is to only pick cards from decks C and D.

As we said, healthy subjects tend to sample cards from each deck at first, but soon realize the correct strategy and pick only from decks C and D. Patients with orbitofrontal lesions do not act this way. They, too, begin sampling all four decks, but never end up focusing on the correct ones. If anything, they stick to decks A and B for the short-term gain, though, in the long run, that method causes them to lose much more than they gain.

Damasio's patient, Elliot, always described himself as a conservative person, cautious with his choices and seldom taking risks.[15] He said this about himself even *after* the operation on his brain. Yet "cautious" would certainly not describe his choices during the gambling task. And it is not as if he didn't understand how to play. He knew his objective. He understood the ideas of gain and loss. He was even able to tell the experimenter which were the good decks and which were the bad ones. Nevertheless, every time he was given the task, he ended up making the worst choices, just like other patients with his type of brain damage.

Why do the patients with orbitofrontal lesions do so poorly compared to the controls? According to Damasio, their difficulty in making decisions is caused by their compromised somatic-marker system. What should happen normally, he says, is that when the subject takes a big loss from decks A and B, he should become much less likely to draw from these decks again. Somatic markers associating those two decks with loss and frustration and anger should be produced and reactivated any time the subject considers picking from them. This process takes place smoothly in healthy subjects because the emotional parts of their brains are in order. In the case of these patients, lesions in their orbitofrontal cortices interfere with the operation of somatic-marker system, infringing on their ability to make prudent choices.

Damasio hypothesizes that when Elliot had tissue from his frontal cortex removed and Phineas Gage suffered the passage of an iron rod through his brain, the resulting damage in both cases hindered the functioning of somatic markers. This is why Gage was no longer Gage and also why Elliot was no longer Elliot. Their abilities to decide and to act in socially acceptable ways based on their years of life experience were severely affected. They no longer could look at the world with the same perspective. Their decisions could no longer be guided by experience. Their opinions, ideas, and behaviors were wildly altered. They were no longer themselves.

Damasio's theory is not proven, but it seems to be founded on astute observations of the brain's organization and human behavior. What we must settle on, then, is whether we accept his hypothesis. But before we decide too quickly, let us take a moment to consider what it implies about human existence.

One tenet of Damasio's theory is that somatic markers can influence our decisions even when we are not aware of it. Before you even begin to evaluate your options in depth, the activity of these markers has already been triggered and your field of choices already narrowed. When you think that you are coming to a decision with an impartial mind, you are actually filled with unconscious biases. When you think you made a decision based solely on weighing the pros and cons, your weighing method was slanted by the operation of somatic markers. You have no conscious control over the production or activation of these markers, although they have control over you, and you may never realize what influence they have.

The operation of these markers threatens our capacity to freely contemplate our decisions as moral agents. Damasio's theory seems to imply that all the choices that we appear to make freely have already been made for us. Does the somatic-marker hypothesis imply that our future is determined by our biological makeup?

No, it does not. Damasio asserts that these markers have a *degree* of control over what we do, but that control is not complete. Somatic markers do not determine our decisions; they just have an influence on them. After the automated triggering of their effects, the conscious self makes the final decision. As Damasio argues, "There is still room for using a cost/benefit analysis and proper deductive competence."[16] The activity of somatic markers may be the initial step, but the conscious self is the ultimate decision maker. This theory, if true, shows that there could be severe restrictions on the power of will, but it still has the final say. As of yet, we can still hold that willful, conscious deliberation is the essence of our daily and moral decisions. However, there does seem to be something faintly disquieting about the notion of a biological process that influences what we do, yet exerts this influence, unbeknownst to the agent, from the depths of the unconscious.

8

THE READINESS POTENTIAL

Suppose that you are investigating the scene of a car accident, hoping to figure out what caused it to happen. The collision occurred on a dead-end street and involved only one car. The vehicle is standing against a wall at the end of the street, its front badly crumpled. You are told by witnesses that the car drove straight into the barrier without slowing down. After inspecting the car, you discover that the braking system is severely damaged. Do you therefore conclude that the accident was caused by defective brakes?

The trouble with this assessment is that you don't know whether the faulty brakes caused the accident or whether the accident damaged the brakes. Which is the cause and which is the effect? The way to sort this out, of course, is to figure out which one happened first. You need to ask the driver if he noticed any problems with his brakes before the collision. If he did not, then the brakes could not have caused the accident. If he did, then the brakes might have caused the accident, though not necessarily. Even if the brakes were defective first, something else could have caused the accident.

What we can learn from this is the simple fact that causes must occur before their effects. Something that happens a year from now cannot cause something that happens tomorrow. We also see another easy concept: an event that precedes another event does not necessarily cause it. To put this in logical terms:

1. A and B are events
2. A comes before B.
3. Therefore, B does not cause A.

This reasoning is effective for disproving a cause-effect relationship. However, as we can see from the example of the car accident, the following argument is faulty:

1. A and B are events.
2. A comes before B.
3. Therefore, A causes B.

This argument clearly doesn't work because an event that follows another event is not necessarily caused by it.[1] The car accident was not necessarily caused by faulty brakes, even if the brakes were damaged first.

Our discussion of the relationship between neurobiology and conscious will has much to do with the sequence of cause and effect. If it is true that our conscious willing of actions is what causes them to happen, then it must be true that our conscious willing takes place before the brain begins executing an action. However, if we could prove, using some experiment, that the conscious willing of actions occurs *after* the brain begins executing them, that would constitute a serious challenge to moral agency. It would mean, as we showed above, that conscious will does not cause our actions. It would mean that free will, and of course moral responsibility, do not exist.[2]

This brings us to physiologist Benjamin Libet, one of a select few scientists who has tried to experiment with human consciousness.[3] Specifically, Libet is known for describing a delay between the body's detection of a stimulus and our conscious awareness of it. He found that when he stimulated subjects' brains, it took them about 500 milliseconds of continuous stimulation to become consciously aware of it.[4] Libet concluded that in order to have a conscious sensory experience, the appropriate brain activities must continue for half a second or longer; otherwise, there will only be an unconscious response to the stimulus. So consider a competitive runner at the start of a 100-meter sprint. When the gun goes off, he will take only about 130 milliseconds to leave his block.[5] This means, according to Libet's results, that the runner has begun the race before he is consciously aware that the gun has fired.

Does the runner freely will himself to leave the block? If it is true that the runner is not aware that the gun has fired when he leaves the block (and he did not cheat by beginning to move early), then it would seem that he does not use his free will. Actually, this is not so surprising because the runner is probably so trained to leave the block at the gunshot that it has become a reflex. He no longer needs to think about it. It is as automatic for him as the retraction of his hand from a hot surface.

Is this a challenge to the concept of will? No. As interesting as this delay may be, reflexes are clearly different from willed actions. We do not consciously control our reflexes. In contrast, willed actions require conscious initiation before the brain can execute them, or at least that is the way it seems to us. What Libet finds in his next experiment, however, is that the way our will seems to work may not be the way it actually works.

It was discovered by scientists Hans Kornhuber and Lüder Deecke in 1965 that each voluntary action is preceded by an increase in the brain's electrical activity. Electrical activity in the brain can be measured using instruments such as an electroencephalogram (EEG), which records the activity using electrodes on the scalp and plots the information as waves. Using this kind of device, Kornhuber and Deecke monitored the brain waves of subjects as they voluntarily flexed their wrists. After filtering the results, they observed a spike in the wave pattern that appeared about 800 milliseconds before each wrist flexion. Agreeing that this spike must be the beginning of the brain's execution of the movement, Kornhuber and Deecke called it the *readiness potential.*

Libet decided that he wanted to extend Kornhuber and Deecke's experiment to discover whether our conscious willing of an act truly causes it to happen, as we feel that it does, or whether it is just some by-product of brain activity. To do so, just like damaged brakes and a car accident, he would have to find out which happened first: the conscious decision to act or the brain's execution of the action. Libet already knew how to identify the latter of the two—that is represented by the readiness potential—but how would he identify the precise time when the subject consciously decided to act? A person's consciousness is not something we can observe from the outside. It is subjective. Recognizing this, Libet took the only option available: he had the subject report the time when he or she made the decision.

Here's how the experiment went. As in Kornhuber and Deecke's experiment, Libet chose to use wrist flexions as the voluntary act. So after placing electrodes connected to an EEG on the subject's scalp, he asked that she flex her wrist whenever she felt like doing so. Libet didn't want to require the movement at certain times because he wanted the subject to act solely based on her decision, without feeling coerced.

The subject was seated in front of a special millisecond clock. Instead of hour and minute hands, it displayed a single spot of light going around and around, completing the circle every 2,560 milliseconds (2.56 seconds). Libet asked the subject to mark the clock time corresponding to the moment she makes the conscious decision to flex her wrist.[6]

After many subjects and many trials of the experiment and after the data was filtered and averaged, Libet obtained a result: the conscious decision to flex the wrist was made about 350 milliseconds *after* the appearance of the readiness potential.[7] *The readiness potential happened first.* The conscious decision happened second.

Libet concluded that the brain begins executing actions about 350 milliseconds before we consciously initiate them. The implication of this is clear: conscious will does not cause our actions. Follow the logic:

1. Free will exists only if we consciously control our actions.
2. If we consciously control our actions, then the conscious decision must occur before the brain begins executing them.
3. The conscious decision occurs after the brain begins executing them.
4. We do not consciously control our actions.
5. Free will does not exist.

According to Libet, free will cannot be the cause of our actions because causes have to occur before their effects. He has shown that the brain activity comes first. Libet cannot, however, conclude that the brain activity causes the conscious decision. That would be like saying that faulty brakes caused the car accident. If event A comes before event B, A does not necessarily cause B.

Interestingly enough, despite his own conclusion that conscious will does not cause our actions, Libet does not deny free will. He believes that it does exist, but that it works differently than we think it does.

Consider Libet's finding that the conscious decision is made 350 milliseconds after the brain begins executing the wrist movement. The decision is also made about 150 milliseconds *before* the actual movement of the wrist (see diagram). Of those 150 milliseconds, the final 50 represent the

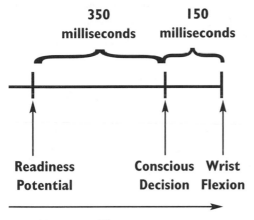

350 milliseconds   **150 milliseconds**

Readiness Potential   **Conscious Decision**   **Wrist Flexion**

Time

Timing Voluntary Action. Libet found that the readiness potential appears about 350 milliseconds before a person consciously decides to act, leaving 150 milliseconds between the decision and the wrist flexion.

time it takes for the brain to activate the muscles in the wrist. Libet says that the remaining tenth of a second is when free will finally comes in. Obviously, it is too late to consciously initiate the decision since the brain has already begun to execute it. Rather, Libet says that the only power of the will is the ability to "veto" an action that has already been unconsciously initiated, just like the president might veto a bill proposed by Congress.[8] Possible actions are processed and initiated unconsciously, and then sent to the conscious self for approval or rejection. After the brain has unconsciously done all the work, all we do consciously is say yes or no. This is Libet's view of free will. As he puts it:

> We may view voluntary acts as beginning with unconscious initiatives being "burbled up" by the brain. The conscious will would then select which of these initiatives may go forward to an action, or which ones to veto and abort so no motor act appears.[9]

There is some evidence for this theory. Sometimes, during experiments, a readiness potential will be detected but no action will follow.[10] In these cases, it could be that the subject vetoed the action that his or her brain had begun to execute.

Damasio's claim was that unconscious processes narrow our field of choices and influence our decisions. Libet is saying that the neurons in the brain are responsible for the entire decision-making process except for the final stamp of approval, which is left to the moral agent. According to Libet, free actions are not free because we consciously will them, but because we have a 100-millisecond window to cancel them before they happen. As one scientist puts it, "This suggests that our conscious minds may not have free will, but rather 'free won't!'"[11]

Libet writes that his view fits with the common understanding of moral behavior:

> This kind of role for free will is actually in accord with commonly held religious and ethical scriptures. Most religious philosophies hold individuals responsible for their actions and advocate that you "control your actions." Most of the Ten Commandments are "do not" orders.[12]

He goes on to write that his experimental results reveal problems with certain kinds of religious beliefs:

> The unconscious appearance of an intention to act could not be controlled consciously.... Therefore, a religious system that castigates an individual for simply having a mental intention or impulse to do something unacceptable, even when this is not acted out, would create a physiologically insurmountable moral and psychological difficulty. Indeed, insistence on regarding an unacceptable urge to act as sinful, even when no act ensues, would make virtually all individuals sinners.[13]

The above two passages appear to be Libet's attempt to demonstrate that his view of the will is not as bizarre as it initially seems. Personally, I find his effort to relate his results to the arrangement of the Ten Commandments and the nature of sin to be a bit silly. No matter how he tries to package it, his view is peculiar, but that alone doesn't mean that it's wrong. If we want to criticize Libet's view, we will have to aim our attack at the substance of his ideas.

Such attacks are not hard to find. Libet's theory is extremely controversial and scores of critiques have been published against it. Generally, these objections fall into two categories: those that dispute the concept of the conscious veto and those that dispute the design of the experiment or interpretation of the results. We will consider each in turn.

The critique of the conscious veto is a simple one: how do we know that the veto doesn't have an unconscious origin just like every other part of the decision? Philosopher Daniel Dennett writes:

> Libet tacitly presupposes that *you* can't start thinking seriously about whether to veto something until you're conscious of what it is that you might want to veto.... But why couldn't *you* have been thinking ("unconsciously") about whether to veto *Flick!* Ever since *you* decided ("unconsciously") to flick, half a second ago? [Emphasis in the original][14]

Libet himself admits this difficulty.[15] If everything else about our decisions is unconscious, why would this veto power be any different?

Dennett's reply, though seemingly valid, makes the situation even worse for moral agency. Libet hypothesizes that we have a conscious veto power. It's not the free will that we'd like to have, but at least he grants us something. On the other hand, Dennett's subtraction of the veto power leaves us with no free will at all. He would have us believe that all our actions are determined by unconscious brain processes over which we have no willful control. With the veto power in doubt, it would seem that the only way to save free will is to find flaws in the experiment itself.

One difficulty that has been raised is whether Libet calculated the time of the readiness potential accurately.[16] The 350 milliseconds that he arrived at was the average of many brain wave times. In some trials, the readiness potential actually appeared after the recorded time of the decision. *On average*, the readiness potential happened first. But what if that average was pulled back by low-hanging outliers? There were certainly some waves that happened way before the others owing to some unrelated brain activities. Perhaps these measurements made the average time earlier than it should have been.[17] After all, the difference is only a matter of a third of a second.

Another challenge to Libet's experiment questions whether the time of the decision was properly recorded. It may have been affected by what might be called *memory bias*, which refers to the way the brain sometimes misrepresents the position of moving objects. For example, in one exercise, subjects were shown a circle that moved across a screen in one direction and then disappeared. Then they were asked to mark the point on the screen at which the circle vanished. It turned out that the subjects overestimated the final position of the circle. Its motion caused them to think that it moved farther than it actually did.[18]

Libet's experiment also involved a moving circle: the one on the millisecond clock. Each subject used this device to mark the time of his or her conscious decision to act. What if they were fooled by the same illusion? It's possible that they, too, believed that the dot was farther along than it actually was, and they recorded a decision time later than they should have. This error in recording the time could explain why the average time of the conscious decision came after the average time of the readiness potential.

Both of the critiques we mentioned have to do with mistakes that could have happened during data collection or analysis. What if the time difference was caused by the design of the experiment itself? Several scientists conducted a functional magnetic resonance imaging (fMRI) study based on Libet's experiment. Rather than trying to interpret brain waves, they wanted to see a picture of the subjects' brain activity as they recorded the time of their conscious decisions. They discovered that as the subjects concentrated on correctly marking the clock time (rather than on deciding to flex), there was an upsurge of activity in several brain areas.[19] This enhanced activity took place before the conscious decision to flex, as the subject was preparing to follow the experimenter's instructions. This early rise in activity could have affected the recording of the readiness potential (which, recall, is calculated by averaging several waves) by making the averaged wave appear to have occurred earlier.[20]

This last critique falls closest to what I believe is the most powerful objection against Libet's conclusion. Rather than questioning the experiment, this argument questions one of its foundational assumptions: that the readiness potential (if not vetoed) leads to the action being carried out.[21] How does Libet know that? The assumption is based on the finding of Kornhuber and Deecke that the readiness potential always appears before voluntary actions occur. Libet effectively concludes that since the readiness potential comes before the action, it must cause the action. But is that a valid argument? Recall the reasoning that we discussed earlier:

1. A and B are events.
2. A comes before B.
3. Therefore, A causes B.

Libet's understanding of the readiness potential could be expressed just this way:

1. The readiness potential and the action are events.
2. The readiness potential comes before the action.
3. Therefore, the readiness potential causes the action.

However, as we established from the beginning, this reasoning is faulty. It would be just like saying that if a car's brakes were defective before an accident, the brakes must have caused the accident. That conclusion does not follow.

The readiness potential certainly could have something to do with voluntary action. It probably does. There is activity in the brain all the time, especially when we are considering how and when to act. The readiness potential could reflect many different brain processes. We can't be certain that it implies that an action has begun to be executed in the brain. A tiny blip in a brain wave plot is no reason to annul free will and moral responsibility. Our deepest convictions tell us that it is our personal introspection and willful execution of thoughts and actions that engender our behavior. It is these convictions that constitute our sense of self and empower our role as moral agents. If we were to accept Libet's ideas, we would have to concede their most ominous consequence: that consciousness deceives us.

9

THE GRAND ILLUSION

"Observe that the coin is in my left palm," the magician tells you. He places a quarter in his left hand, leaving his right hand open to reveal that it is empty. His left palm now clenched into a fist, the magician brandishes his wand and, with a bellow of magical words, taps the wand to his fist. Slowly, he opens his hand. "What do you see?" he asks.

The coin has vanished. Your senses tell you that the magician caused the coin to disappear with a flick of his wand. For a brief moment, it seems even to be the clearest explanation for the sequence of events. You know, however, that this explanation cannot be true. Despite your perception that the wand caused the coin to vanish, you are certain that wands cannot do such things. You are left to assume that what you perceived was an illusion.

You are, of course, correct. The explanation is more complex than the magician would have you believe. Hidden in his hand was a thin elastic cord with an adhesive end. The cord extended from the magician's hand, through his sleeve, to the inside of his jacket. When he placed the quarter in his hand, he stuck it to the end of this string and clenched his fist. At the climax of the illusion, the magician slowly opened his hand, releasing the tension on the elastic cord and allowing the coin to whisk into his sleeve. The coin was transported by the use of props you were unaware of. The wand and the magic words were an act. The magician used them only to maintain the illusion.

Daniel Wegner is a psychologist who believes that we are continually being fooled by this kind of illusion. The illusion is that of conscious will and the magician is the brain.

In short, Wegner denies that conscious will exists. He acknowledges that each of us has a certain conscious experience when we initiate voluntary actions. As you raise your hand to ask a question during a lecture, you have the ineffable sense that your conscious wish, the command of *raise your hand!* that is going through your head, is what is causing you to raise it. Wegner does not deny that you have this feeling. He denies that this feeling is what causes the action. He writes:

> The fact is, it seems to each of us that we have conscious will. It seems we have selves. It seems we have minds. It seems we are agents. It seems we cause what we do.... It is sobering and ultimately accurate to call all this an illusion.[1]

Just as the magician creates the illusion that the movement of the wand causes the coin to disappear, the brain creates the illusion that the mental command *raise your hand!* causes your hand to raise. In each case, you (the audience), witness clever showmanship. The magic words, the feeling of deciding, the movement of the wand, the experience of willing—these events do not cause the end result. They are just to maintain the illusion. Beneath the magician's theatrical performance lies a simple, mechanical explanation. The laws of physics tell us that the taut elastic cord in the magician's jacket stores enough potential energy to pull the coin out of his hand once the tension is released. Wegner believes that the illusion of conscious will can also be dispelled by realizing that, underlying the decorative feelings of will created by the brain, there is also a raw, mechanical explanation. It involves the electrical activity of neurons as they exchange bits of information throughout the brain, and the synthesis and breakdown of chemicals and molecules that occurs in every inch of brain tissue. However, like the elastic cord, we don't see any of this. All we experience is the conscious feeling and then we see the result: the hand is raised. And just as we might, even for the briefest moment, believe that the wand has caused the coin to vanish, we continually give in to the illusion that our thoughts can cause our actions.

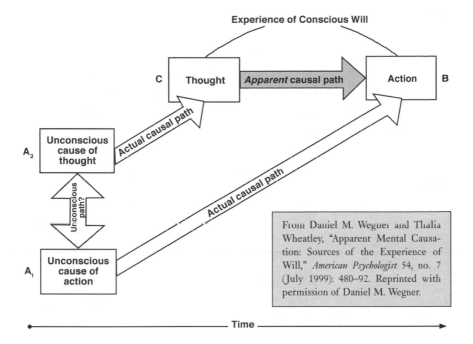

From Daniel M. Wegner and Thalia Wheatley, "Apparent Mental Causation: Sources of the Experience of Will," *American Psychologist* 54, no. 7 (July 1999): 480–92. Reprinted with permission of Daniel M. Wegner.

Wegner writes that there are three factors that help maintain the illusion of conscious will. The first of these, which he calls *priority*, is simply that the feeling of mental willing comes before the action. This, as we found in the previous chapter, is certainly crucial because causes have to come before their effects. If someone were to sense that her conscious willing of an action came after the action already happened, she would of course conclude that the action was unwilled.

The next factor he cites is *consistency*, the fact that what we consciously will ourselves to do tends to match what we end up doing. If it does not, the experience of will is lost. If you will yourself to scratch your chin and you end up picking your nose, you will deny responsibility for the nose pick because you will consider it an unwilled action. The content of the feeling must be consistent with the action.

Wegner's third criterion is *exclusivity*, the idea that the illusion of thought causing action is maintained as long as we cannot identify obvious alternative causes. The feeling of exclusivity might be lost in a situation where someone hears voices in his head, experiences motor tics (as in Tourette's), is electrically stimulated to move by a deranged neurosurgeon, or is told to do something while hypnotized. Wegner claims that as long as these three criteria are present, a person will perceive that his or her actions are willed, even if they aren't.[2]

Imagine that you are sitting by a window on a windless day, looking outside at a tree.[3] You look at one of the branches intently and picture it moving up and down. Most surprisingly, you discover that as you do this, the branch outside actually moves in this way. Puzzled by this curiosity, you picture a different branch moving in a circle, and behold: it does just that. Now this is getting very weird. You picture a branch making a figure eight, and it does. You picture the branches spelling out your name, and they do. Suddenly, you run over to your neighbor's house and tell everyone that you can consciously will the tree to do the Macarena.

In this semi-ridiculous example, Wegner claims that the feeling that you control the movement of the tree branches would result from the presence of his three factors: the branches move only after you picture them doing so, their movement is consistent with the movement you imagine, and there aren't any other events (that you can see) that could be causing the branches to move as they are. Thus, you have the impression that your free will can control the tree. This, however, could be an illusion (you're probably on drugs). If you find it hard to relate to this example, that's probably because it is very, very weird.[4] The question is, can this thought experiment be made into a real experiment?

It can. In fact, the experiment has already been done by Wegner and his colleagues.[5] In what he called the "I-Spy" study, Wegner sat a subject in front of a computer screen displaying a number of different objects. During each trial, the subject was instructed to move the mouse to the object named in a recorded message. For example, if the recording said "swan," he would take the mouse and move the cursor to the position of the swan on the screen. What he is not told, however, is that he has no control whatsoever over the movement of the cursor. His movement of the mouse does not affect its position. The cursor is actually being controlled by one of the experimenters. Nevertheless, the subject still believes that he is the one moving it. Wegner says that this happens because the three factors (priority, consistency, and exclusivity) are present. That is enough to convince the subject that he is using his free will. That's all it takes to create the illusion.

This experiment, of course, is not proof that free will does not exist. It only tells us that we can sometimes be tricked into thinking that we have free will when we do not—but we already knew that. It's a very old idea that traces back to philosopher René Descartes' writings in the seventeenth century. Descartes held that we can't ever be completely sure that we are in control of our actions because there is always the possibility that we are being fooled. After all, we could be dreaming.

> How often have I dreamt that I was in these familiar circumstances, that I was dressed, and occupied this place by the fire, when I was lying undressed in bed? At the present moment, however, I certainly look upon this paper with eyes wide awake; the head which I now move is not asleep; I extend this hand consciously and with express purpose, and I perceive it; the occurrences in sleep are not so distinct as all this. But I cannot forget that at other times I have been deceived in sleep by similar illusions; and, attentively considering those cases, I perceive so clearly that there exist no certain marks by which the state of waking can ever be distinguished from sleep, that I feel greatly astonished; and in amazement I almost persuade myself that I am now dreaming.[6]

Wegner's experiment confirms for us the possibility raised by Descartes: our senses can be fooled. This, however, is not enough to demonstrate that we *are* being fooled. His central claim still needs to be justified. So to give substance to his theory, Wegner does what every other theorist in this debate does: he turns to the brain.

If it is true that conscious will can initiate our bodily actions, Wegner says, then conscious will and the initiation of bodily actions would have to be part of the same system in the brain. There must be some neurobiological chain of events that begins with the conscious initiation of an action and ends with that action being carried out. However, what if it could be demonstrated that such a connection does not exist? What if it could be shown that the feeling of will and the execution of behavior are completely independent processes?

"The motor structures underlying action are distinct from the structures that allow the experience of will," Wegner says. "The experience of will may be manufactured by the interconnected operation of multiple brain systems, and these do not seem to be the same as the systems that yield action."[7] Wegner seeks to invalidate the view that conscious will causes action by showing that action and the feeling of will are not part of the same cause-effect pathway in the brain. If this is so, it might mean that the connection we experience between thought and action is, in reality, an illusion.

Scientists know a little bit about the functions of the various regions of the brain. The execution of bodily movements, for example, is generally traced to the motor cortex, a stripe of brain tissue that runs across the center of the brain from ear to ear. Let's consider this to be a major component of the "system that yields action," to use Wegner's words. Of course, it is likely that almost the entire brain plays some role in yielding action, but we will pretend that it all comes down to the motor cortex, at least for the moment.[8]

Similarly, we will pretend for the moment that the experience of will comes from the frontal lobe. Nobody actually knows where the experience comes from—in fact, figuring this out could be considered the paramount challenge of neuroscience—but we will choose the frontal lobe simply because the region has been cited as the source of relevant abilities such as executive function.

It could be said that, because the experience and the action might arise from two different areas of the brain, they must be unconnected processes. However, this assessment certainly can't be correct and it can't be what Wegner is saying. There is no reason why different brain regions cannot cooperate to achieve a result. That is how the brain works. Take vision, for example. After frequencies of light are received by the eyes, in the front of the brain, they are processed in the occipital lobe, all the way in the back of the brain. Despite these different locations, the visual pathway can still be considered one system.

What Wegner must be saying is this: wherever they may be located in the brain, the systems for the feeling of will and the initiation of action cannot be *causally* connected. The "conscious will region" does not cause the activity of the "action-yielding region." Is this true biologically? Nobody knows. Because of the limits of scientific knowledge about the brain, it appears that this argument cannot proceed any further on the neurobiological front. However, Wegner believes that the separation between the experience of will and the execution of action may be apparent in human behavior.

A person afflicted with alien hand syndrome might find his alien hand (whether it be a tentacle, claw, or otherwise) involuntarily unbuttoning his shirt. He in no way wills his hand to do this—in fact, he will probably try to stop it—but the action takes place nonetheless. The action proceeds without any conscious initiation.

When electrical impulses are directed at the motor area of someone's brain, parts of his or her body can be induced to move without the person's conscious consent. In one such experiment, a neurosurgeon used electrical stimulation to cause a patient to move a hand and then to speak. In response to the movement, the patient said: "I didn't do that. You did." After being induced to speak, the patient said: "I didn't make that sound. You pulled it out of me."[9] These are simple demonstrations, but it should be possible to use the same method to cause more complex unwilled acts. In fact, if the proper stimuli were applied by some wild, drunken neurosurgeon, a person could probably be electrically induced to do the Macarena.[10]

The point is that the behaviors that people typically associate with conscious will can happen without ever asking consciousness for permission. Actions can happen without the feeling of will. This is an example of, as Wegner puts it, the action and the experience of will "coming apart."[11] The illusion of free will stems from the perceived connection between the feeling of willing an action and the action itself. Examples such as these, Wegner says, demonstrate that the feeling and the action may not be as connected as we think. However, we need not go as far as electrical stimulations and alien hand syndrome to understand that unwilled actions can occur. Every time a doctor tests a patient's reflexes with a hammer, he is inducing an action that the patient has no conscious control over. A person might not will herself to retract her hand from a hot stovetop, but that movement occurs automatically. Do these examples indicate a gap in the connection between behavior and conscious will? Probably not. We all accept the fact that there are both willed and unwilled actions. It should come as no surprise that there are bodily movements that we don't mentally initiate. Anyone will also acknowledge that conscious initiation of an action is not the *only* way to cause it. I can mentally will my knee to bend, but a doctor can also cause my knee to bend with his hammer. The knee movement is an effect that can occur as a result of various causes. Conscious will is a possible cause; it isn't the only possibility.

So what are we to say about Wegner's proposition that the experience of will and the execution are completely separate processes? Since the evidence from unwilled actions is thin, there isn't much holding up the theory at this point. However, there is another way to consider the separation. Rather than showing that actions can occur without the experience of will (since we know this already), a more persuasive example would be to show that one can have the experience of will without the action—the feeling that an action was consciously controlled when, in reality, the action never took place.

The case that comes to mind is that of phantom limb syndrome. Most patients who have had a limb amputated experience a mystifying phenomenon: they continue to experience the presence of the limb after it has been removed. There have been cases in which the illusion lasted for twenty-five years after the amputation.[12] Some amputees report feeling tingling sensations in their phantom limb. Others experience pain.[13] There are those, however, who experience a much stronger illusion.

In some cases, amputees have reported experiencing movement of phantom limbs—not twitches or spasms, but complex, voluntary movements. When the patient wills the invisible limb to move, he experiences it responding precisely according to his conscious command. He experiences details like the wiggling of his fingers and the bending of his elbows and knees. The movement of each part feels consistent with the movements of the others; none seem to be moving abnormally. The realization that the limb is not really there does not make the feeling go away. All the subtle features of movement combine to create the inescapable impression that he is truly moving his arm or leg.[14]

Amputees who suffer from this syndrome have the experience of consciously willing an action even though that action doesn't actually take place. This is an example of a situation in which the feeling of will and the yielding of action come apart. The feeling is there; the action is not.

Phantom limb syndrome is not the only example of this kind of separation between the feeling of will and the action. It happens sometimes to patients with schizophrenia.[15] They might feel as if their thoughts can control the movement of things in their environment, like that of the sun across the sky or trees in the wind. As with phantom limb syndrome, these are cases in which a person has the experience of conscious will despite the fact that the action supposedly willed never actually happens. They lend support to the idea that the feeling of will and the execution of action are independent processes.

Although it depends on where you work, it isn't every day that you come in contact with schizophrenics or people with phantom limbs. If it is true that the feeling of will is independent of the initiation of action, then we should be able to find signs of this division in everyday life. Descartes' example of dreaming might be one of them. I can recall many instances when I had the experience of willing myself to get out of bed and brush my teeth, only to wake up seconds later and realize that I was still in bed and no such action was taken.

There are, however, more subtle examples than this. The feeling of will comes and goes all the time. It may not appear and disappear as sharply as it does in the examples we discussed, but the strength of the feeling can be reduced to very low levels, even for actions that we would take for granted as being willful.

For example, the first time you play a song on the piano, it takes a lot of mental focus. Each key press feels willed. Every fingering, every press of the pedal, every note and chord feels as if it is being strictly controlled. With practice, the song becomes easier to play, and less concentration is needed. You stop thinking about the individual notes and begin to appreciate the melody. At the same time, the feeling of will begins to fade. You no longer focus on individual notes. You no longer feel that you are mentally commanding each press of the key. In fact, you might feel as if your fingers are moving faster than your thoughts. The sense of control is gone. It is as if your fingers are playing by themselves.

We could say the same about you as you are walking home. Being an experienced walker, you tend not to pay attention to the precise movements of your feet. You have no conscious experience of initiating each step. Rather, you are thinking about what you did that day or why the weather is so crummy or the fact that you need a haircut. You are not consciously lifting each leg, extending it, planting it on the ground, and then pushing the other one forward. Your legs are moving more or less automatically. However, if I asked you to walk, with one foot in front of the other, along a straight line, you would probably pay more attention to your steps. You would direct the movements of your feet more carefully and exert mental effort into ensuring that they line up at each step. The feeling of will would return.

One moment you feel as if you are willing the action, the next moment the feeling is gone. Then it returns. Wegner writes:

> This, in turn, suggests the interesting possibility that conscious will is an add-on, an experience that has its own origins and consequences. The experience of will may not be very firmly connected to the processes that produce action, in that whatever creates the experience of will may function in a way that is only loosely coupled with the mechanisms that yield action.[16]

One of the most powerful examples of this loose coupling of the feeling of will and the yielding of action (and, yes, this will be the last example) is that of hypnosis. When a person is hypnotized, she is put into a kind of sleeplike trance in which she is aware but much more responsive to suggestions. It is said that when the hypnotized person acts on the suggestion of the hypnotist, she feels as though she is acting freely, but her behavior is actually determined.[17]

In one experiment, the subject was instructed to open a window upon hearing the word "Germany." When the subject heard the word, he made up a sensible reason for why he needed to open the window, saying something like "It is awfully stuffy in here, we need some fresh air. Do you mind if I open the window?"[18] The subject was convinced that he made the decision freely and because of his own reasons, but this is an illusion. Wegner believes that this illusion supports the existence of the grand illusion: that the feeling of conscious will can control our behavior.

If we accept Wegner's view, we know the implication: the end of free will is also the end of moral agency. We cannot be considered morally responsible for our actions unless we can freely control them. Wegner does his best to piece together some idea of moral responsibility that is consistent with his theory. He writes that conscious will is "the person's guide to his or her own moral responsibility."[19] What this translates to is that the feeling of will is just part of the human machine. It is the basis for moral emotions that are part of the causal chain that determines our actions. The feelings of authorship we each have over our actions are illusory, yes, but Wegner believes that they are "the building blocks of human psychology and social life."[20] But as far as the truth of the matter goes, moral responsibility cannot exist in Wegner's world. It is just another part of the grand illusion.

The lingering question is whether Wegner is correct. Of course, we cannot prove beyond a reasonable doubt whether he is right or wrong, but we *can* evaluate his position by the evidence he brings to the table.

Wegner's view rests on the division that he claims exists between the feeling of will and the yielding of action. He cites many examples to support this separation, but none of them appear to be that convincing. We already discussed why instances of unwilled actions like those of alien hand syndrome or electrically induced movement don't imply anything about free will. Wegner would have us believe that these behaviors indicate that the system that generates the feeling of will and the system that generates action are independent from one another, but we don't have to reach this conclusion. Why should it be surprising that arms or legs can move without being willed to do so? The fact that there is more than one way to cause an action doesn't mean that conscious will cannot control the action. Alien hand syndrome and electrically induced movements don't tell us any more about the will than do knee reflexes induced by a rubber hammer. It is true that many human actions are done without the involvement of free will, but that doesn't mean that free will doesn't exist.

If examples of actions without will aren't convincing enough, what about those of will without actions? Patients with phantom limb syndrome have the experience, which can feel completely authentic, of being able to control a limb that isn't there. Since the feeling of will is there without any action taking place at all, could it be that the feeling and the execution of the action are separate? Again, I don't think that this is the most reasonable conclusion. The fact that we can be tricked about whether we are exerting willful control over one action does not necessarily mean that we are being tricked all the time. The conscious mind can play all sorts of tricks. Optical illusions, for example, fool us into thinking we are seeing something that isn't really there. We can be deceived into thinking that straight lines are crooked, congruent shapes are incongruent, or shapes exists where they do not. Does that mean that our perceptions are always wrong? Does that mean that the system that processes frequencies of light is independent of the system that creates perception? No. The scientific view at this time is that these systems are completely intertwined. Wegner has not eliminated the possibility that the systems for the feeling of will and yielding action are connected in the same way.

The last of Wegner's ideas that we discussed was the fact that the intensity of the experience of will can fluctuate. In a trancelike state, such as during hypnosis, the intensity of the feeling can be high, even though the person is barely in control of her actions—if she has any control at all. While playing the piano or walking down the street, the feeling of will can fade or even disappear, even though we would call those things willful acts. Wegner insightfully points out that the experience of will can fade and intensify, but should this lead us to conclude that it is a feeling separate and independent from the control of behavior? That conclusion simply does not follow from the evidence. Rather, the explanation might be that the body can enter certain states in which our conscious control becomes more limited and, therefore, our feeling of control becomes weaker while we are in those states. It cannot be implied by the mere fact that the intensity of a feeling can change that the feeling itself does not parallel our actual ability to control our actions.

The trouble with Wegner's attempt to prove that conscious will is an illusion lies in the kind of evidence he provides. Though his examples are thought provoking, they can't tell us anything conclusive about the will. They can be interpreted in countless ways and are not developed enough to show that conscious will is an illusion like that of the magician's disappearing coin. To prove that the mind is not the cause of behavior, Wegner would need a stronger kind of evidence: controlled, scientific experimentation. The question is, what kind of experiment would he need? How could we demonstrate that the will does not exist? To say that conscious will is an illusion is also to say that human actions are determined. We also know that, like the path of a stone rolling down a hill, determined actions are in theory predictable. If we could find a way to use the neuronal information in a person's brain to predict his or her behavior, that could be the foundation for a powerful argument against free will and moral agency. If all our actions are predictable, that could mean that they are determined. If our actions are determined, then it must be true, as Wegner claims, that the experience of willing is an illusion.

10

NEURONAL DESTINY

The total eclipse of the sun by the moon is one of the most spectacular natural phenomena one can witness. In ancient China, each eclipse was considered a heavenly omen, one foreshadowing the life of the emperor.[1] To the Chaldeans it meant that people had incited the moon's fury and that great misfortune was to come.[2] In the seventh century BCE, the poet Archilochus wrote of the eclipse:

> Nothing there is beyond hope,
> Nothing that can be sworn impossible,
> Nothing wonderful, since Zeus,
> Father of the Olympians,
> Made night from mid-day,
> Hiding the light of the shining sun,
> And sore fear came upon men.[3]

Two hundred years later, the Greek historian Herodotus wrote an account of an eclipse that once took place during a battle:

> As...the balance had not inclined in favour of either nation, another combat took place in the sixth year, in the course of which, just as the

battle was growing warm, day was on a sudden changed into night. This event had been foretold by Thales, the Milesian, who forewarned the Ionians of it, fixing for it the very year in which it actually took place. The Medes and Lydians, when they observed the change, ceased fighting, and were alike anxious to have terms of peace agreed on.[4]

The tone of Herodotus seems different from that of Archilochus, who sees the eclipse as astounding, extraordinary, and even ominous. He wrote that it was an act of Zeus that brought "sore fear" upon men. Herodotus, on the other hand, does not seem either perturbed or amazed by this rare orientation of the heavenly bodies. Rather, he said that the event was foretold. It was expected to happen.

What happened during those two hundred years of history that changed the perception of the eclipse so drastically? The answer is scientific realization. In the sixth century, Thales of Miletus, the founder of Greek geometry and astronomy, used the scientific principles he had derived to figure out when the next solar eclipse would occur. On May 28 of the year 585 BCE, the prediction of Thales came true.

What Thales realized was that, despite all the surrounding mythology, the eclipse was not a miraculous act of Zeus but a natural, determined phenomenon. As such, the timing of the eclipse would have to follow natural laws. Thales also understood that if one has access to enough information, natural phenomena can be predicted.

Consciousness, decision making, intention, behavior—these are also natural phenomena that have been surrounded by mythology. Yet it is generally accepted that they are not events beyond the physical realm. They do not break the laws of nature. Shouldn't that imply that they, too, are determined? As we have seen, there are many who believe this to be the case. As it happens, there is actually a way to demonstrate this. If it is truly the case that our intentions, decisions, behaviors, and the like are determined, then, as the accomplishment of Thales shows, we should be able to predict them—provided we have access to enough information. We of course do not know everything about the brain, but neither did Thales know everything about the nature of astrophysics and the cosmos. If our decisions are as inevitable as the next solar eclipse, then the current techniques and instruments of neurology should allow us, at least on a simple level, to ask neurons about our destiny.

In 1927, Ivan Pavlov conducted an experiment that would be cele-brated ever since by scientists in many disciplines. Pavlov observed that his dog salivated whenever he delivered food to it. You might say that he was the deliverer who brought salivation, but that would just be a bad joke. Anyway, curious about this salivation, Pavlov decided to train his dog by consistently ringing a bell before bringing it food.[5] He continued this training for some time. As is well known, Pavlov's famous result was that, eventually, the dog began to salivate at the sound of the bell.

Pavlov's training conditioned the dog to expect food at the sound of the bell. The release of saliva occurred because the dog's nervous system announced that food would soon be ingested, and the digestive system began preparing by causing saliva release. The salivation took place *in anticipation* of eating food.

The brain doesn't drool, but it does generate anticipatory signals in a similar way. Using neuroimaging technology, for example, it is feasible to monitor the areas of the brain that are activated when a person engages in a certain behavior or thinks a certain thought, such as to see activation in the amygdala during emotional processing or activation of the motor cortex during voluntary movements. This is the typical use of brain scanning tech-niques like functional magnetic resonance imaging (fMRI) and positron emission tomography (PET). We have encountered such applications of neuroimaging before. Some scientists, however, have proposed that moni-toring brain activity could be used for a radically new purpose: *predicting* how a person will behave. These researchers view human decision making just as Thales viewed the eclipse, as a deterministic natural phenomenon, and, like Thales, they seek to demonstrate this through prediction.

The idea is to find patterns of brain activity that anticipate behaviors, much as the dog's salivation takes place in anticipation of its eating. A number of neuroscientists, for example, have tried to interpret brain scans to predict bodily movements. One of them is Apostolos Georgopoulos.

Georgopoulos inserted electrodes into the motor cortex of a rhesus monkey to record the neuronal activity there.[6] He then sat the monkey in front of a lever that could be pushed or pulled into eight different posi-tions. The monkey was trained, using rewards of food, to move the lever in a direction indicated by an illuminated light at the target position. The electrodes in the monkey's brain recorded the firings of the neurons that were activated during this action. Georgopoulos then repeated this proce-

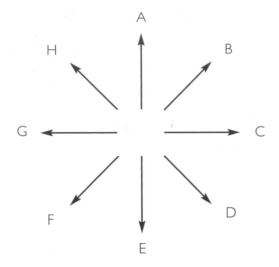

Neuronal groups for each lever direction

Georgopoulos found that neuronal group A in the monkey's brain was activated when it moved the lever up, neuronal group E was activated when it moved the lever down, and so on.

dure for each of the other seven directions of the lever, recording the pattern of neuronal firing for each one. The diagram on this page shows the eight lever directions, each label corresponding to the firing of a certain group of neurons (represented by capital letters).

The way neurons code for the direction of movement is surprisingly similar to the way we vote for a political candidate. The result of the presidential election is determined by adding up the votes of every citizen (let's ignore the Electoral College, since the brain doesn't work that way). Each person gets one vote, and the winner of the election is the one for whom most people vote for.

Each neuron in the motor cortex can be thought of as voting for a direction of movement. Each neuron gets one vote and the resulting direction of movement is the sum of all the votes. The difference between this process and the election, of course, is that there are an infinite number of directions as opposed to the two or three political candidates. However, the process still works.

Let's suppose that we are studying a brain with only two neurons. If both neurons code for the "up" direction, the movement will be up. If they both code for "right," the movement will be right. But what if the first neuron codes for "up" and the second codes for "right"? The result of this neuronal election will be the sum of the two:

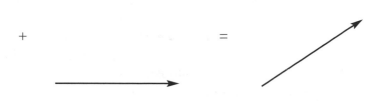

This process works the same way when millions of neurons are voting. The resulting motion is equal to the sum of the directions coded by every neuron in the voting pool.[7]

When Georgopoulos recorded the brain activity corresponding to each movement of the lever, he figured out which neurons were coding for which direction. He found that group A neurons coded for the "straight up" direction, group B coded for the "up and to the right" direction, and so on. He found out, to use our earlier analogy, the voting intentions of the neurons he was recording from.[8] Knowing the precise intentions of the voters, whether they be neurons or American citizens, is enough to predict the results of the election. And that is exactly what Georgopoulos did.

In a new phase of the experiment, Georgopoulos did not signal the monkey to move the lever in a specific direction. Rather, he instructed it to move the lever in *any* direction. The electrode recordings told Georgopoulos which neuronal groups were being activated, and did so about 300 milliseconds before the monkey's hand began to move.[9] Since he knew how each of those neuronal groups liked to vote, Georgopoulos was able to predict the direction in which the monkey would move the lever.

Without question this is an incredible scientific advancement. Its applications are numerous and may include the development of prosthetic limbs that can be mentally controlled. The basic idea is that if you know that the activation of a certain neuronal group (we'll call it group X) causes the right leg to step forward, then, in theory, you could construct a prosthetic right leg that steps forward whenever neuron group X is activated. Scientists at Duke and Cambridge universities, for example, recorded neuronal activity as a monkey controlled a robotic arm using a joystick. The scientists then disconnected the joystick and wired the robotic arm directly into the monkey's brain. After some practice, the monkey was able to control the robotic arm with its thoughts.[10] A similar study at Brown University succeeded in allowing a human being to mentally control electrical devices plugged into his brain. The subject, who happened to be a quadriplegic, was able to move a cursor on a computer screen, check his e-mail, flip though channels on his TV, and play a computer game—using only his mind.[11] These are all extensions of Georgopoulos's work.

The question, however, is what his finding tells us about free will. On the surface, the implication is clearly against free will, but there are two difficulties with the experiment that we could cite in defense of free will. The first is that Georgopoulos used monkeys in his experiment. Whether or not monkeys have free will is debatable. Most people would probably agree that monkeys do not have moral agency. For example, if your pet monkey Coco steals a cart full of candy from the grocery store, nobody would consider Coco responsible. *You* would be blamed, even if you denied that you paid him to do it.

The second difficulty is that the study is too limited. The experiment shows us that neuron firings cause bodily movements. But so what? We knew that already. It takes some time for the signal to move from the brain and arrive at the muscles in the arm. That just happens to give us the time to say that a signal is headed toward the muscle and that a movement is imminent. The possibility is not eliminated that the neuron firings that sent the signal were the result of a conscious decision to move. Just the fact that the muscles are stimulated by neuronal signals doesn't imply that consciousness is not involved.

Perhaps the prediction of bodily movements isn't the right kind of study for our investigation of free will. The main demonstration of free will that we have discussed is decision making. It seems reasonable to say that the making and carrying out of a decision is the prime example of free will in action. Has any experiment succeeded in foretelling the result of a conscious choice—one made by a human being? The answer is yes. Using neuroimaging techniques, scientists have in fact been able to predict the decisions of human subjects.

A group of neuroscientists at Washington University's School of Medicine in St. Louis designed a computer game that they asked a number of volunteers to play.[12] The objective in this game was to look at a group of moving dots and detect the direction in which they were moving. The subjects would indicate the direction of the dots by pressing one of four keys (up, down, left, right).

The group of dots could appear anywhere on the screen and would only stay visible for one-fifth of a second, making it quite easy to miss. So eleven seconds before the start of the trial, the volunteers were given a hint: an arrow was shown briefly on the screen, pointing to the region where the dots were likely to appear. This hint gave the correct direction 80 percent of the time. The rest of the time it pointed to the wrong part of the screen. It was up to the volunteers to decide whether they would use the hint or ignore it.

While the subjects were playing the game, their brain activity was monitored using functional MRI (fMRI). This technique measures activity in a region of the brain by detecting the amount of blood flowing to that region. The greater the blood flow, the greater the activity. The researchers discovered that different patterns of blood flow (brain activity) reflected whether or not the volunteers used the hint.

As we said, the game involved an eleven-second gap between the appearance of the hint arrow and the moving dots. In that time, by looking at the pattern of blood flow in the subject's brain, the researchers were able to determine with high accuracy whether or not the subject would offer a correct answer. As one researcher put it, "Before we present the task, we can use brain activity to predict with about 70% accuracy whether the subject will provide a correct or incorrect response."[13]

These neuroscientists have been successful in predicting the performance of volunteers playing the game. They were able to foretell the results of a human decision. What they were unable to do, however, was predict the content of those decisions. The researchers were completely unaware of the thoughts, the conscious deliberation, that went into those decisions. If the workings of the mind are as determined as the time of the next solar eclipse, then our thoughts should be predictable as well.

Nobody has yet been able to predict human thoughts, but there has been some success in monitoring them—what some might call neurological *mind reading*. Neuroscientists, for example, have been able to match certain thoughts to specific brain waves on an EEG (electroencephalogram). With this information at hand, a researcher can hook someone up to an EEG with a net of electrodes, monitor her brain activity, and know what she is thinking when those specific brain waves appear. Use of the EEG in this way is valuable not only to our discussion but also to criminal investigations.

The standard lie detector, the polygraph, works by detecting changes in bodily states. Changes in heart rate, breathing, blood pressure, and sweating that occur while a subject is being interrogated are used as signs that he or she is lying. This often works on people who are nervous that the questioning will reveal their deception. However, there are also many cases in which the polygraph detects a lie when the person is actually telling the truth. This probably happens because the innocent person is simply nervous about being strapped to a polygraph. There are also people who are pathological and can easily lie without getting nervous. They can sneak their fabrications past the polygraph undetected.

Researchers have been able to overcome these problems by developing a new kind of instrument: a brain-based lie detector. One such technology, called *brain fingerprinting*, is thought to have enormous potential for use in criminal investigations. In fact, it has already been used to provide evidence in court.[14] During a brain fingerprinting test, the subject wears a helmet lined with electrodes. The helmet is connected to an EEG to monitor brain activity. While the subject's brain waves are being recorded, the interrogator shows him words or pictures. Among the things presented, there will be words or pictures that relate to the crime being investigated. If these words or pictures are unfamiliar to the subject, the EEG data will show nothing of interest, just normal waves. But, if the subject recognizes one of words or images—if he is surprised by it or finds it significant—a specific wave called a *P300* (positive wave that appears 300 milliseconds after the onset of a stimulus) will be recorded by the EEG.[15]

Suppose that the person being questioned is a suspect in a homicide case. The prosecutor believes that he shot his friend with a certain 9 mm pistol at an old barn. The suspect denies having ever seen this barn or ever having touched a handgun in his life. During a brain fingerprinting test, interrogators observe a P300 wave when a picture of the murder weapon and the old barn are shown. The prosecutor now knows that the suspect lied. He recognizes both the barn and the murder weapon, and this can be used as evidence against him. The accuracy of this method is almost 100 percent.[16] No matter how good of a liar the suspect may be, he can't hide what's going on in his brain. There's nothing he can do to prevent the interrogators from reading his mind.

For many people, the three scientific demonstrations in this chapter have great implications for our self-understanding. If simple movements and decisions can be predicted, that means, sooner or later, when technology and scientific understanding are more advanced, scientists will be able to predict all our actions. What's more, if scientists can make simple discoveries about what we are thinking, such as whether we recognize something, then we should expect that sometime in the future they will be able to predict all our thoughts. These experiments, some will assert, represent the initial scientific evidence that free will is an illusion—that what we believe to be moral agency is actually just the determined operation of a network of neurons.

This would be both an amazing and a terrifying conclusion, if correct, but I don't think it is true. The reason, I think, has to do with the nature of the behaviors that were predicted in these studies. We first discussed bodily movements. Scientists have been able to use neuronal activity to predict the motion of an arm or leg. We then looked at the prediction of a simple human choice. That subject had to decide on the direction of a group of dots, and do so based on a hint given earlier (an arrow pointing to the probable location of the dots on the screen). Finally, we looked at a simple case of what some might call neurological mind reading. Scientists have been able to detect whether a person considers a word or picture to be surprising or significant.

But none of these are true moral decisions. A moral dilemma involves a conflict between ideas that cannot be reconciled using any formula or clear set of guidelines. A moral decision draws upon personal experiences, desires, and needs. It is guided by broad principles of what is good and what isn't—principles that cannot simply be taken at face value because they depend on the situation at hand. They must be interpreted based on context. The decider might have to take into account both ends and means, with the understanding that some consequences cannot be prepared for. During the crucial phase of mental deliberation, the person might call upon his every conscious power—emotion, reason, memory, intention, creativity, self-reflection—and undergo difficult inner struggle, exerting his mind and pushing the limits of his thinking, until finally he makes his decision and wills himself to act.

This is not to say that all moral decisions are that dramatic. Moral decisions are also not the only ones that require serious conscious deliberation. All I want to highlight here is how involved a human decision can be.

Now consider again Georgopoulos's experiment with monkeys pulling levers. Using neurological techniques, he was able to predict the movement of a limb. First of all, I am not sure that simple bodily movements are good examples of consciously willed acts. I might even be willing to grant that many of the movements we make are determined. Sometimes, when I listen to music I like, I tap one of my feet to the rhythm. I am often unaware that I am doing it and don't remember having done it. My foot tapping in that case may well be a determined action. My breathing also generally happens automatically. There are, of course, plenty of movements that require willful control, but we could all agree that there are some that don't. However, this issue is not the main problem I have with the interpretation of the study.

Let us assume that the movement being studied *does* require conscious deliberation. How can we be sure? Georgopoulos asks a person to participate in the study and to move the lever. The person puts a lot of thought into the movement. He makes the decision of how to move the lever intently, basing it on various things from his experience. After a long thought process, he decides that since his favorite heavenly body is the North Star and he often takes business trips to the North Pole, he will move the lever straight up. The decision is not random, but made carefully and with intention. Could Georgopoulos predict *that* movement?

I believe he could, not because I am abandoning free will or moral agency, but because of the constraints of Georgopoulos's experiment. Recall that Georgopoulos was able to make his prediction about 300 milliseconds before the monkey actually pushed or pulled the lever. Recall also what this decision was based on: the firing of neurons in the motor cortex. This is the same way it would work with the person who loves the North Star and the North Pole. When the subject decides to move the lever forward, he consciously initiates the action. This causes a chain of neuronal events that eventually leads to the activation of neurons in the motor cortex. By observing the activation of these neurons, Georgopoulos can predict the movement.

It turns out that Georgopoulos is not predicting the subject's decision at all. He has no clue about the person's decision-making process. There happens to be a small delay between the moment at which the brain receives instructions for a movement and the moment at which the movement actually happens. All Georgopoulos does is look at the neuronal activity during this period to figure out what movement the person decided to initiate.[17] He looks at the *result* of the person's thinking. *The conscious decision was already made.* The result of the decision just happens to appear slightly earlier in the brain than it does in the arms or legs. It takes time for the message to travel from the brain to the arms or legs. Georgopoulos takes advantage of this to get a sneak preview of the movement on a computer screen before anyone else can see the subject move. He cannot predict the movement an hour before it happens. He cannot predict how or what the person will decide.

The second experiment we discussed also uses a kind of trick to predict a subject's performance in a game. If you remember, the object of this game is to press one of four buttons to indicate the direction in which a group of dots is moving. Eleven seconds before those dots appear, the subject is given a hint: an arrow that points to the area on the screen where those dots will probably appear. The hint is accurate 80 percent of the time and misleading the other 20 percent. The researchers found that different patterns of brain activation result depending on whether the subject uses the hint. They also know when the hint will be helpful and when it will be misleading. Using these two bits of information, they can, before the trial is over, say whether the subject will be right or wrong.

Has freedom of the will been undermined? Again, I don't think so. As is the case with Georgopoulos's experiment, this one does not actually predict a decision. The subject has four choices of buttons to press: up, down, left, or right. The experiment doesn't predict which button he will press, only if he will be right or wrong. The researchers have no way of knowing how the subject is deciding. What they see is whether or not the hint is accurate and whether or not the subject uses it. For example, let's say that the hint is accurate and brain patterns reveal that the subject uses it. What's the prediction? The subject's response will probably be correct. If the hint is misleading and the subject uses it, the response will probably be incorrect. This is not a true prediction of what the subject will decide; it is a loophole.

Imagine that you were the subject in the experiment, but instead of being a typical participant, you have a secret agenda: you want to ensure that the scientists fail to predict your performance. You could do it easily. When you see the hint, and later the dots, you pay careful attention. On those trials when the hint is accurate, the researchers will assume that you are using the hint properly and will get the correct answer. You, however, enter the wrong answer intentionally. If the scientists were truly able to predict your decisions, they would be aware of your sneaky plan, but they don't actually have that ability. They might as well guess what your choices will be.

The two studies of behavior prediction that we have looked at both depend on recognizing patterns of activation in the brain. The first one uses activation of neurons in motor cortex, an event that occurs after the decision is made. The second one makes a prediction based on a pattern of neuronal firings that clearly does not determine the final decision, since the person can easily trick the experimenter into predicting incorrectly. It is clear that those brain activations do not *determine* the behavior of the subjects as the laws of physics determine the moment of the next solar eclipse. Neither of these studies has been able to gain access to human thoughts and *really* predict a decision.

Brain activation is also used in the lie detection technique brain fingerprinting, which some might consider a type of mind reading. I'm sure we can agree that being able to discover whether a person recognizes a picture is a long way away from predicting human thoughts. The question we need to answer for all of these scientific developments, however, is whether they will lead to real methods of neurological fortune telling. Are these kinds of experiments bound to fail when they try to take on the core of human consciousness? Or, are we witnessing the beginning of the end of moral agency?

A growing belief is that the latter is true. More and more, brain and mind are rising to the forefront of scientific research. New drugs and chemicals are modifying the operation of the brain. New technologies are promising to change the way we think. A simple pill can be taken to sharpen our minds, improve our social life, or just make us more effective at handling life's challenges. Perhaps what we call free will or agency is merely the mechanical operation of the brain. Perhaps all human behavior is subject to the deterministic rules of neuronal interaction, and, in the end, our decisions are predictable. Today we might think that consciousness is a thing of mystery and that conscious will is free, but we still can envision the possibility that the strategy of Thales of Miletus to predict the behavior of heavenly bodies could someday be applied to predict the behavior of human beings. What would we say then?

11

THE REVOLUTION OF THE BRAIN

In his book *Listening to Prozac*, psychiatrist Peter D. Kramer talks about a patient of his named Sam, an architect by profession, who had a strong liking for something that he realized would be frowned upon by most people: pornography. He had a genuine interest in sex films and was not in the least bit embarrassed to admit it. In fact, he was proud of his perspective on the matter, believing that he was promoting sexual individuality. His wife had very much the opposite opinion, having no interest in pornography whatsoever. Sam thought that she was too uptight to appreciate the videos that he liked so much.

After experiencing problems with his business and the death of his parents, Sam was overcome by depression. He came to Dr. Kramer for help. At the time, a new treatment for depression had just been introduced to the pharmacology market: a drug called Prozac. Few doctors had prescribed this novel compound and Kramer was unsure about giving it to patients, but Sam agreed to try it out.

The result was astonishing. Sam reported not only that he was feeling better but that he was "better than well." His depression vanished. His memory and focus improved. Even his work habits and public speaking style were sharpened. But despite all the benefits that Prozac provided him,

Sam felt that something was wrong. He was concerned because he lost his enthusiasm for pornography. His emblematic belief, his distinguishing little idiosyncrasy, had always been his fondness for a more open sexual lifestyle. Sam was distressed by the fact that this special aspect of his personality was suddenly deleted when he began taking Prozac.[1]

Throughout his book, Kramer discusses the way Prozac, in many cases, not only treated patients of their depression, but modified their sense of self. He wonders about the disturbing fact that human personality can be so altered by the ingestion of a tiny pill. In a similar vein, we might wonder whether that fact compromises our ability to control our behavior as moral agents. These questions concern not only Prozac, but the myriad of chemical technologies available today that can manipulate the way we think.

The way antidepressants like Prozac usually work is by shifting the balance of chemicals in the brain. The main chemical that tends to be targeted by these drugs is the neurotransmitter *serotonin*. Neurotransmitters are compounds that allow neurons to communicate with one another by helping to carry the message. Because of its connection to antidepressants, serotonin is probably the most famous example of such compounds. We might think of serotonin as the "happy neurotransmitter." There are actually at least two others that could be given that title, namely, *dopamine* and *norepinephrine*, both of which we will discuss later. As for serotonin, however, the basic trend is that, up to a certain point, the more of it you have, the happier you are. The less of it you have, the less happy you are.

The original purpose of antidepressants and similar drugs was to treat people with depression that was so severe as to be considered an illness. People with strong suicidal tendencies, for example, would qualify. One would think that to fall into the category of needing drugs to fix someone's day-to-day behavior, that person must exhibit tremendous emotional instability as well as an equally tremendous chemical imbalance in the brain. However, with the explosion of new drugs and related chemical products on the market, the bounds of what we consider to be illness are rapidly widening. What once might have been considered a heavy-hearted personality, or perhaps the romantic melancholy of the great poets, is now termed "depression," and there are drugs made to treat it. Today's psychiatric drug market has turned into what Kramer calls "cosmetic psychopharmacology."[2] The term designates the rise of a new trend to take drugs not because of a serious mental syndrome but to boost one's mental abilities. People want to make themselves, like Kramer's patient Sam, "better than well." Just as one might use cosmetics to enhance physical attractiveness, some people are using, and being encouraged to use, chemical compounds to enhance the performance of their minds.

Take the case of the drug Provigil, named by combining the words in the phrase "promotes vigilance."[3] The drug was originally designed for the treatment of narcolepsy, an uncommon condition that has as its symptoms inadequate sleep at night and uncontrolled bouts of sleepiness during the day. We all have trouble sleeping sometimes and often we all feel drowsy during the day. That is not narcolepsy—not even close. People suffering from the disorder often fall into deep sleep while standing in line or while engaged in conversation. Most alarming, unless the person is diagnosed and warned not to drive, he might spontaneously fall asleep while at the wheel of a car. These incidents may happen at any time, without warning, and the person might wake up as quickly as he fell asleep, often feeling dizzy and bewildered.

Provigil acts against the drowsiness of narcolepsy and closely related disorders without any side effects.[4] It doesn't cause your hands to tremble or heart to race as excessive caffeine might. Effective against narcolepsy, the drug just keeps you awake and keenly aware of what goes on around you.

It has been estimated that narcolepsy affects around 1 out of every 100,000 people.[5] In the United States, only about fourteen new cases are diagnosed each year.[6] Why is it, then, that Provigil sales brought in $1.2 billion in 2005? The reason is that most of the people using it didn't have narcolepsy at all. They were perfectly healthy people taking it as a "lifestyle drug" to augment their day-to-day efficiency. People feeling a bit tired took it to make their fatigue disappear. People who felt that they weren't thinking clearly took Provigil to sharpen their minds and get back in the game. But if the drug can do all these things, why stop there?[7]

Consider how dangerous it would be if an airline pilot wasn't feeling too awake or clearheaded as he sat in front of the controls. How useful would it be if we could simply give him a pill to sharpen his mind? In research studies of the drug, Provigil kept pilots awake and aware for nearly three consecutive days.[8] Imagine what this could do for the average person. Imagine how human productivity could increase if we could go days without sleep while maintaining our alertness and level of accomplishment. Research is being conducted to see whether the drug could be used to make truck drivers more aware on the road overnight, help workers in a night shift stay energetic, and empower soldiers to undertake prolonged combat operations with added vigor and intensity.[9] In fact, the drug was used (under a different brand name) by French soldiers during the first Gulf War.[10]

Provigil is not the only drug with this kind of potential. Research has shown that the drug Donepezil, developed to treat Alzheimer's disease and other serious forms of dementia, was shown to make pilots *better at flying planes*. Eighteen pilots were divided into two groups and tested on a sophisticated flight simulator. The simulator was an accurate representation of the cockpit of a common aircraft model called the Cessna 172 and generated a very similar flight experience. Initially, both groups performed equally well. Then, each group was given a bottle of medication and instructed to take 5 milligrams a day for 30 days. Unknown to the pilots, one group was given a bottle of Donepezil, the other was given placebos. After 30 days passed, the pilots each got into the flight simulator again. The result was clear: the pilots who had taken Donepezil performed better in the flight simulator. The drug had enhanced their skills.

You don't have to have a job as a pilot or truck driver to find a use for brain-enhancing drugs. College students all over the country have discovered the advantages of abusing the attention deficit disorder drug Ritalin to help them succeed in school. Students claim that the drug keeps them focused and allows them to do their work more effectively. In a 2001 study of students at 119 universities, it was found that about 7 percent of them abuse some kind of prescription stimulant like Ritalin. The rate of usage tends to be higher at more competitive colleges.[11]

Aside from augmenting academic or professional performance, the major application of cosmetic psychopharmacology is probably the enhancement of social interaction. The definition of what is normal social behavior can vary a great deal, but the advent of the pharmacological revolution has begun to set the standard. Years ago, no one would stop and consider whether a shy person has some psychiatric syndrome. No one would look at a quiet person who enjoys spending time alone and wonder what was wrong with her. However, the explosion of new drugs on the market now provides a "cure" for what might be called *social phobia*. What years ago might have been thought of as a reserved personality is now considered outside the social norm. It is a condition that lends itself to medication.[12]

To be sure, there are people who truly do have serious psychiatric or neurological disorders that impair their ability to get along with other people. They might need medication; there is no doubt of that. But there are also perfectly healthy people who use these drugs cosmetically, to make themselves more social or just to help them fit in. Just as Provigil is used to increase vigor and alertness, there are people who take individual drugs or even combinations of drugs to achieve social grace.

Taking antidepressants like Prozac is one approach to instantly boost your social skills. In his book, Peter Kramer describes how a patient of his, Tess, exhibited this effect while on the drug. When Tess came to see him, she told Kramer about all the problems in her personal life. Her relationships with men were dysfunctional. She didn't know how to interact with them properly, and they tended to stay away from her. Her social life was far from vibrant. In addition, Kramer found that Tess had the symptoms of depression. He decided to put her on Prozac.

Just like Sam (the guy who liked pornography), Tess found herself feeling "better than well." Not only did her depression go away, but she suddenly possessed a newfound set of social skills. The talent for dealing with men that she always wanted was suddenly instilled in her. She came into Kramer's office boasting of multiple dates in a single weekend. Tess ditched her old group of friends, deciding she couldn't relate to them anymore, and found a group of new ones. Socializing was effortless for her. Without having been given any tips or advice on relationships, Tess's social life blossomed quickly—all thanks to chemical effects of antidepressants in her brain. Tess herself recognized this. She once told Kramer: "I call myself Ms. Prozac."[13]

Besides antidepressants, another chemical way to obtain social grace, shockingly enough, is to use a class of antipsychotic drugs called *benzodiazepines*. These drugs, when used in small doses, have been shown to improve social ability.[14] The number of things that someone might find wrong with his or her social life is vast, but then again so is the availability of drugs. Klonopin, Xanax, Wellbutrin, Neurontin, Effexor, Lamictal, Aricept, Keppra, Luvox, Zoloft—the list goes on. There is enough out there that anyone who wants to chemically elevate his or her social aptitude can find some drug out there to do the trick.

Perhaps you are comfortable with your social skills. There is still much about your mind that you can upgrade. I'll bet your memory isn't perfect.

There may be ways to improve that, too, aside from all the drugs meant for Alzheimer's disease. The potential exists for a pill that prevents us from forgetting. Nobel Prize–winning neuroscientist Eric Kandel and his research group were able to elicit this kind of effect in mice. In the brains of mice and human beings, there are molecules that allow for memory formation, such as cyclic AMP (cyclic adenosine monophosphate), which regulates neuronal messaging processes crucial to sustaining memory. By encouraging the presence and continued activity of cyclic AMP and related molecules, a scientist can enhance a person's, or a rat's, recollection. This is precisely what Kandel's team did. They constructed a maze for mice to navigate and trained them to use it. After the mice had forgotten how to get through the maze, Kandel and his colleagues gave the mice a drug to encourage the activity of cyclic AMP. Suddenly, the mice were able to navigate the maze again, thanks to the memory-enhancing effects of the drug.[15]

This same principal could be applied to human beings to improve our memory. With this kind of technology, forgetting an event could become a thing of the past, unless of course one *wants* to forget that event. We all have memories that we find unpleasant or even traumatizing—things we would rather not remember. If a drug could be developed that improves memories, in principal, it should also be possible to create a drug that deletes memories. We might be able to wipe the mental slate clean of anything unpleasant.

The cosmetic use of pharmaceutical products for the mind is spreading. As one psychologist says, the utilization of chemical compounds to improve mental abilities and behavior may one day be as "ordinary as a cup of coffee."[16] At the rate we are going, the day may come when every person you meet is medicated in one way or another. Competition in the workplace and just about every challenge in life might be met with a new prescription. A neurologist writes: "One plausible scenario is that neurologists will become quality-of-life consultants. Following the model of financial consultants, we could offer a menu of options with the likely outcomes and risks.... Prospecting for better brains may be the new gold rush."[17]

Neurologist Richard Restak, whose book *The New Brain* talks at length about how society and technology affect brain development, finds the uses of psychiatric treatments we've discussed strangely unsettling. He is concerned not only because these uses of prescription drugs have gone beyond the bounds of medicine but also because they appear to imply certain things about the nature of human beings. The fact that people are taking drugs in order to improve their personalities or their mental functions sounds eerily close to the concept of upgrading a computer. It seems that the pharmacology market considers the human being to be a machine—a computer-like mechanism composed of organic flesh instead of circuitry.[18] Restak puts the problem this way:

> Herein lies the conundrum: If we think of ourselves as little more than chemical machines that can be altered by drugs, then what happens to traditional concepts like free will and personal responsibility?... What will be the overall result... people interpreting their experience in chemical terms rather than interpersonal ones?[19]

Our behavior is treated like computer output. If the output (behavior) is undesirable, inject a few chemicals to fix the mechanism. Feeling blue? Bored? Why should you? You can repair that annoying symptom of your brain chemistry with a wide range of pharmaceutical products. Human emotions, what many of us consider so central to our identity as moral agents, are no longer being regarded as the special longings of the human spirit, but as a measure of brain functioning. Being depressed no longer indicates an unfulfilled desire or missed opportunity, but a shortage of serotonin, dopamine, or norepinephrine in the brain. Pride and sorrow, ecstasy and shame, fury and passion—these are just products of the human machine. Much of the respect that once existed for human consciousness is lost.

Restak describes the kind of patient who exhibits this perception of a human being and who treats himself accordingly. His name is Ted. Because his job requires him to travel often and switch time zones on many of his trips, Ted uses Provigil to keep his sleepiness under control. Before one of his trips, Ted's brother, Jim, died in a car accident. Ted was deeply saddened by the news, but he didn't want to be. He took a drug with tranquilizing effects to make his melancholy go away. He was emotionless at his brother's funeral. Then he went back to work.

Ted told Dr. Restak that he began thinking about his brother again two weeks later and that he burst into tears "for no reason." Dissatisfied with this "morbid" behavior, Ted decided that he must have some sort of depression and that antidepressants were the solution. Ted remembered having used them several years before. He took them during a period when he said he was questioning the things in his life, trying to figure out his purpose in the world. The antidepressants made these curiosities fade away, replacing them with a feeling of contentment that such drugs tend to provide. Ted thought the drugs would help him stabilize himself in his current situation by ridding him of his grief. Combined with his use of Provigil to keep him alert during the day and a sleeping pill to give him peace at night, Ted would establish for himself a medicated existence that would allow him to function normally each day, without the interference of emotion.[20]

The view that the mind is a mechanism that should be managed by chemical treatment is spreading. One by one, our mental abilities are being selected as new targets for the pharmaceutical industry. More and more aspects of consciousness are being considered adjustable and due for enhancement. Before the modern revolution of the brain, which I believe is just beginning, the road to self-improvement consisted of personal reflection, advice from mentors, considerations of one's goals and values, and a commitment to learning from one's mistakes. At present, a trend has begun that threatens to replace that approach with a neurobiological one. A growing number of people are adopting a new perception of the self that calls for all issues of personality to be mended mechanically, using the right chemical compounds—whether those include Prozac, Provigil, or otherwise. The view holds that by modifying the interactions of neurons, you can repair any of the flaws of personhood.

It cannot be denied that developments in pharmacology have transformed the medical treatment of disease. There is no doubt that they have given hope to millions who years ago had little. But have they silenced the moral agent?

Moral agency is founded on the assumption that our thoughts and actions are not determined, not generated by the mechanistic action of neurons or neurotransmitters. It is our conscious will, inspired by our experiences and emotions, our contemplation and deliberation, that directs the engine of brain mechanics to guide us through life. This seemingly natural understanding of the person is threatened by the rising tendency to represent the human being as a programmed machine, of which all thoughts and actions are merely outputs, results of the deterministic interactions occurring within. If what now exists only as a supposition, based on recent pharmaceutical innovations, evolves into an established worldview, we will be forced not only to say that our everyday actions are determined but also our most treasured moral decisions. It will have to be concluded that our deepest moral convictions are not actually "ours" at all. They belong to the brain, to neurons, to chemical compounds—to the vast chain of causes tracing back to the beginning of time. The implications of such a worldview for the workings of society could be tremendous. If deterministic chemical exchanges in the brain are the sole cause of our behavior, then we cannot justly be held responsible for immoral actions—the inclination to do what's ethically wrong cannot possibly be overcome by the moral agent. The source of human evil is the brain.

12

SEEDS OF CORRUPTION

It is May 21, 1924. In Chicago, at around five o' clock in the evening, two friends named Richard Loeb and Nathan Leopold are driving toward the Harvard School, a prestigious college-preparatory school for children of wealthy families.[1] Loeb, eighteen years old, is a brilliant, handsome graduate of the University of Michigan, the youngest graduate in the history of the school. Soon to enter Harvard Law School, Loeb is the son of a former vice president of Sears, Roebuck and Company and has access to as much money as he could ever want.[2] Leopold, too, comes from an affluent Chicago family. His father is a millionaire who made his fortune manufacturing boxes. At nineteen years old, he is a wonderfully intelligent student of law at the University of Chicago and fluent in at least five languages.[3] A lover of ornithology, Leopold has given lectures on the topic and is a leading expert on a rare bird called the Kirtland's warbler.[4] Everyone expects great things from Leopold and Loeb, certain that they are young men of promise. What no one expects, however, is that these young men are on their way to commit what they believe will be the perfect crime.

Leopold and Loeb have been planning to kidnap someone—anyone, it doesn't matter who—collect the ransom money, and kill him or her. As they are driving to the Harvard School, they spot that person. It happens to be their mutual acquaintance, fourteen-year-old Bobby Franks. Loeb calls the boy over. He asks Bobby to get in the car so they can discuss a tennis racket. Bobby does so, and in no time the car speeds off. As Leopold drives, Loeb stuffs a cloth in Bobby's mouth and strikes him in the head several times with a chisel.[5]

After murdering Bobby Franks, Leopold and Loeb go looking for a place to hide the body. They decide to leave it in a ditch beneath the train tracks, assuming that no one would ever find it there, and begin to write their ransom note, which demands $10,000 from the Franks family.[6] They never receive that money because Bobby's body is found very quickly and, before long, investigators trace the crime to them. The boys confess.

The news of the cold-blooded murder dominates every media outlet, not only in Chicago, but throughout the country. The nation is both appalled and confounded by the incident. Who would have thought that Leopold and Loeb—brilliant young men from elite families, with bright futures, blessed with every advantage—could have committed such an atrocity? For what possible reason? It's not as if they needed the money.

Even more consuming than the news of the murder is the trial of Leopold and Loeb, an impassioned convocation mobbed with fiery opinions, malignant rumors, and frenzied rioters. The public demands that the killers be hanged for their crime. And they probably would have been, had it not been for Clarence Darrow, their defense lawyer.[7] Darrow is the one who decides that the boys should plead guilty before the court.[8] Most people expect that they will plead not guilty by virtue of insanity (this probably wouldn't have worked, especially since it was clear from their elite education that they knew right from wrong), but Darrow is prepared to defend them in a different way, one unprecedented in the history of murder trials.

He claims that the act of murder by Leopold and Loeb was *determined*. In a beautifully composed twelve-hour-long oration, Clarence Darrow explains to the court that The two were caused to kill Bobby Franks by their biological makeup. They were compelled by deterministic forces that they had no power to resist:

> I know, Your Honor, that every atom of life in all this universe is bound up together. I know that a pebble cannot be thrown into the ocean without disturbing every drop of water in the sea. I know that every life

is inextricably mixed and woven with every other life. I know that every influence, conscious and unconscious, acts and reacts on every living organism, and that no one can fix the blame.... Is Dickey Loeb to blame because out of the infinite forces that conspired to form him, the infinite forces that were at work producing him ages before he was born, that because out of these infinite combinations he was born without it? If he is, then there should be a new definition for justice. Is he to blame for what he did not have and never had? Is he to blame that his machine is imperfect?... I know that somewhere in the past that entered into him something missed. It may be defective nerves. It may be a defective heart or liver. It may be defective endocrine glands. I know it is something. I know that nothing happens in this world without a cause.[9]

It was Darrow's appeal to biological determinism, among other arguments, that saved Leopold and Loeb from the death penalty.[10]

Let's take a moment to consider Darrow's strategy in the courtroom. His clients are wealthy and well educated. They are physically and mentally fit, without any trace of neurological illness. Yet they committed a brutal murder. Clarence Darrow's defense is that the act was determined— by neurons in their brains, hormones in their blood, their upbringing. The "machines" of their bodies, their brains, are responsible for the act.

If it's true, however, that we are talking about two apparently sane young men, as everyone involved in the case attested to, then we are left to confront an unsettling possibility. Criminals with mental illness are protected by the insanity defense. That is clear. That leaves criminals without mental illness, who, if found guilty, must take full responsibility for their actions. However, if Darrow's defense worked for Leopold and Loeb, it should work for every criminal. Darrow himself believed this, as evidenced from a speech he gave to the inmates of Cook County jail in Chicago:

> The reason I talk to you on the question of crime, its cause and cure, is because I really do not in the least believe in crime. There is no such thing as crime as the word is generally understood. I do not believe there is any sort of distinction between the real moral condition of the people in and out of jail. One is just as good as the other. The people here can no more help being here than the people outside can avoid being outside. I do not believe that people are in jail because they deserve to be. They are in jail simply because they cannot avoid it on account of circum-

stances which are entirely beyond their control and for which they are in no way responsible.[11]

Darrow implies that every criminal should be able to appeal to the *determinism defense*, which would not only reduce the criminal's moral responsibility but eliminate it altogether. This is because we know, of course, that there can be no moral responsibility without free will, and there cannot be free will if our actions are determined. As two sociologists have put it: "The notion of biological predisposition can relieve personal guilt by implying *compulsion*, an inborn inability to resist specific behaviors."[12]

That assertion represents the core of the determinism defense. It means that people do not freely choose to behave immorally. A life of crime is in their blood, or better yet, *in their brains*.

If we are to decide whether this is true, it is clear where our investigation begins: in the brains of criminals. Since scientists who study this issue focus their investigations on violent offenders—people convicted of murder, assault, and arson—we will do the same. Those crimes are far more interesting than shoplifting and tax evasion anyway.

A number of researchers have examined the composition of cerebrospinal fluid in criminal subjects. Cerebrospinal fluid, often abbreviated CSF, is a clear liquid found between the skull and the cerebral cortex and in the spinal column. Since it is filled with important metabolic proteins, samples of it are often taken to help diagnose neurological disorders. In these studies, however, CSF samples were taken so researchers could look for a correlate of violent behavior in the brain—and they did find one.

One study took CSF samples from thirty-six violent criminals, some of whom killed their victims and others who attempted, but did not succeed, in doing so. The criminals had in common that they were all male, that none of them used guns during their crimes, and that their attacks were especially cruel. The results of the experiment brought to light another commonality: the CSF samples of twenty-seven of the violent offenders had low levels of a compound known as 5-hydroxyindoleacetic acid, or 5-HIAA.

Recall that serotonin is a neurotransmitter that makes us feel content and happy. The compound 5-HIAA is a by-product of serotonin activity. Whenever serotonin is used, some 5-HIAA is generated as a result. Imagine that you are eating individually wrapped candies and I am trying to figure out how many you've eaten. Whenever you eat a candy, a by-product consisting of the empty candy wrapper is formed. The number of candy wrappers left over is a reflection of how many candies you have consumed. In much the same way, the concentration of 5-HIAA in the cerebrospinal fluid is a measure of how much serotonin has been consumed. That means that the less 5-HIAA you have, the lower your serotonin activity, and the less content and happy you must be. That was the conclusion drawn from evaluating these twenty-seven criminals.[13] The low level of 5-HIAA in their brains was an indicator that their bodies were not metabolizing serotonin as much as the average, nonviolent person. This neurotransmitter deficiency, however slight, could have caused these men to be less content and more aggressive. Perhaps it caused to them to be killers.

The other nine criminals in the study did not show any signs of reduced 5-HIAA levels. Their levels were significantly higher than those of the others criminals. This could be meaningless, of course. What we are concerned about is the amount of serotonin being used. A low concentration of 5-HIAA is only one of many things that can go wrong with serotonin. Perhaps these nine criminals had problems with receptors for serotonin, meaning that, though there was enough serotonin around, neurons in the offenders' brains were having trouble recognizing and using it. This is one of several possibilities that could support the claim that their serotonin wasn't working properly. The researchers in the study, however, did not believe this to be the case.

There was something about those nine men that was different from the others in the study: their acts of violence were premeditated. The nine criminals did not kill in sudden fits of rage. Their crimes were planned in advance. Perhaps this was the explanation for their higher 5-HIAA levels. Looking back at the group of twenty-seven criminals, the researchers concluded that their low concentration of 5-HIAA correlates not with violent behavior (since the violent behavior of the other nine didn't correlate with 5-HIAA deficiency) but with *impulsive* violence, or impulsive behavior in general.[14]

In another study, forty-three violent criminals and arsonists were divided into two groups, based on whether their crimes were considered impulsive or not impulsive (the arsonists all fell into the impulsive category). A third group was made up of healthy volunteers who were not criminals. As before, it was found that the impulsive offenders had the lowest 5-HIAA concentrations in their cerebrospinal fluid. The surprising finding was this: the nonimpulsive criminals had the highest 5-HIAA concentrations of all three groups—higher even than those of the healthy subjects.[15] This discovery suggests that serotonin levels are involved in premeditated violence as well as impulsive violence.[16] Support for this claim comes from studies of aggressive behavior in monkeys, in which it has been observed that changes to the activity of serotonin are connected with goal-oriented, aggressive behavior.[17] It may be that that too little serotonin activity leads to impulsive violence while too much activity leads to premeditated violence.

Could it be that serotonin levels in the brain determine criminal violence? That explanation seems exceedingly simplistic. It can't be true that the cause of a complex behavior like aggression, engendered by the most intricate physiological system in the world—the human brain—is that easy to figure out.

As we might have guessed, serotonin is not the only neurotransmitter whose operation (elevated or reduced) has been correlated with aggressive behavior and violence. Another common example is dopamine. Like serotonin, dopamine is involved with the system of generating pleasurable feelings. It helps elicit feelings of joy in order to encourage and reinforce certain activities. For example, the body attempts to reinforce activities that allow for nourishment of the body and continuity of the species by releasing dopamine while a person is eating (probably depending on how good the meal is) or having sex (I'll refrain from commenting). Also like serotonin, low dopamine activity has been shown to contribute to aggressive behavior.[18]

Another relevant neurotransmitter is norepinephrine, which is involved in attention and the fight-or-flight response. Its activity has also been implicated in human aggression and violence. For example, in a study of twenty violent criminals, twenty arsonists, and ten healthy, nonviolent subjects, an association was found between low norepinephrine activity and impulsive aggression.[19]

Could these findings be used to defend convicted criminals in the courtroom? Perhaps they fall into the category of being "not guilty by reason of insanity." The legal history of the insanity defense is complex, so we won't discuss it much, but historically, the rule of thumb to demonstrate insanity has been the M'Naghten rule:

> To establish a defense on the ground of insanity, it must clearly be
> proved that, at the time of the committing of the act, the party accused
> was laboring under such a defect of reason, from disease of the mind, as
> not to know the nature and quality of the act he was doing; or if he did
> know it, that he did not know what he was doing was wrong.[20]

A client's legal insanity has also been settled on in the courtroom by proving that he or she acted because of a unique compulsion or "irresistible impulse," such as kleptomania.[21] We can probably guess that low serotonin levels and kleptomania are not viewed equally by the legal system in the degree to which they reduce responsibility. Nevertheless, the role of neurotransmitters in determining aggressive behavior has been used in court.

In one case, the defendant, we'll call him Bill, was guilty of killing his wife. He did it, yes—but was he morally responsible for the act? Bill had a long history of aggressive behavior. He admitted having been involved in over a hundred fights. Bill did very well in college, always earning impressive grades. Soon after his graduation, he and his wife divorced and the two of them were involved in bitter arguments about child visitation rights. One day, furious and drunk, Bill took a handgun and went to see his ex-wife. His plan was to cause her severe emotional pain by shooting himself in the head in front of her. When Bill arrived, his ex-wife saw him with the gun and, terrified, ran to a neighbor's house. Bill was, to say the least, upset that his plan to kill himself in front of her failed. He followed his ex-wife and, in his drunken fury, murdered her.

After this terrible incident, samples of Bill's cerebrospinal fluid were tested. Though his 5-HIAA levels were normal, it was found that the activity of serotonin and dopamine in his body were low. These findings were introduced by his defense attorneys during trial. The lawyers argued that Bill did not plan to kill his wife in advance. His crime was a result of impulsive aggression, which can be caused, at least in part, by reduced activity of serotonin and dopamine. Since the court did not believe that Bill met the requirements for being not guilty by reason of insanity, the biological observations were dismissed. Bill was sentenced to death.

As we can see from this case, changes in the levels of serotonin, dopamine, and norepinephrine that we have been discussing are typically not recognized as inducing insanity. The concentrations of these neurotransmitters in the body vary from person to person and from one kind of personality to another. So far, convicted criminals with varied levels of these brain chemicals tend not to succeed when using the insanity defense. It seems they will need the determinism defense.

There may be much more evidence for the determinism defense than we have pointed to thus far. It is widely believed, for example, that people fall into a life of crime because of the familial circumstances in which they were raised. Cold, alcoholic parents; poverty; poor education; desperation; early exposure to drugs—these are the social and environmental conditions that many believe lead to a future of robbery, violence, and imprisonment. It's common to presume all of this, but is there any neurobiological evidence of it? Actually, there is.

There is a biological concept known as *hormonal imprinting*. It is the idea that the early interactions between a hormone (many neurotransmitters are hormones) and its corresponding receptor, together with which it yields its effect, will determine their interactions in the future. Early in life, receptors for hormones are in development. If the receptor, during this period of development, has only slight interaction with its hormone, its later activity will be affected in a way that causes its activity to be slight. If the receptor's initial contact with its hormone is intense, the receptor-hormone interaction will be intense in the future. If its early interactions are defective, they will be permanently defective. The lifelong relationship between hormone and receptor can be determined by their interactions when a person is very young, even before the person is born.[22]

What this means is that aggressive and violent behavior exhibited by criminals may have been determined by interactions in their brains *before they learned to take their first steps*. Researchers have shown how it can happen.

Newborn rats treated with endorphins, a natural type of opiate, began acting aggressively during adulthood. The males bit the females and the females kicked the males. Further tests revealed that many of the rats had low serotonin levels, which reduced their level of contentment.[23] When researchers injected another group of young rats with antiserotonin drugs, they found that those rats developed low serotonin levels in their adult life, demonstrating that the early disruption in serotonin activity caused permanent effects.[24] Rats treated with morphine before birth developed abnormal norepinephrine activity in adulthood.[25] Also, early exposure to cocaine has been shown to reduce the activity of dopamine.[26]

What we see here is testimony for the possibility that factors in a person's early life—like exposure to cocaine or any other circumstance that might alter the activity of serotonin, dopamine, or norepinephrine—can determine that he or she will be violent in adulthood. A future of crime may be determined at birth.

And it may be determined even earlier than that. Many studies point to the likelihood that there is a genetic factor at work. The classic culprit is the XYY genotype. As is well known, the X and Y chromosomes are sex chromosomes. Females normally have two X chromosomes while males have an X and a Y chromosome. Some men, however, are born with an extra Y chromosome, giving them an XYY genotype. As a result, some of them develop slightly more masculine characteristics. They tend to be taller and more well built.[27] It has been suggested that these men have extra testosterone, which would increase their tendency toward aggressiveness.[28]

Research studies done in prisons have shown that a disproportionate number of inmates have the XYY genotype.[29] A group of researchers has also discovered that the rate of property crimes is higher among XYY males than typical XY males.[30] With this information in mind, it should not surprise you to learn that connections have been found between having an XYY genotype and having a low concentration of 5-HIAA in the cerebrospinal fluid (a sign of low serotonin activity) as well as having poorly functioning serotonin.[31]

Scientists have discovered another consequence of having the additional Y chromosome. This may come as a surprise to some, while to others it has been obvious all along: those with this extra male chromosome have decreased intellectual capacity. Intelligence tests given to XYY males and control subjects confirmed that the XYY males were of significantly lower intelligence. This actually may be the explanation for the higher crime rate.[32] Being unintelligent doesn't only mean that they are more likely to commit crimes but also that they are more likely to get caught and therefore show up as statistics in criminology studies. In one case, after illegally triggering a fire alarm, an XYY male was easily arrested because he decided to stay where he was—so that he could admire the fire trucks as they arrived.[33]

In fairness to my fellow males, we will need to go after the females for a moment. After all, though men commit 85 percent of all violent crimes in America, women do their share as well.[34] There are certain female attributes that increase the likelihood that women will commit crimes. The major example is premenstrual syndrome, or PMS. Several studies have shown that women engage in violent crime much more often during and right before menstruation.[35] This is understandable, since PMS symptoms include things like short-temperedness, mood swings, and anxiety. But what is the explanation in the brain? It all comes back to serotonin. Changes in serotonin levels, as well as the levels of a number of hormones, are thought to cause the symptoms. Severe PMS is often treated using antidepressants that increase serotonin activity.[36]

In fact, levels of all three neurotransmitters we have discussed—serotonin, dopamine, and norepinephrine—can be adjusted using pharmaceutical treatments. Does this mean that criminality can be cured? Actually, to a degree, it can be. One approach might simply be to prescribe antidepressants, which tend to increase the activity of serotonin. This was tried with a group of twelve criminals, each with a history of aggression. They were given placebos for two weeks, followed by antidepressants for three weeks, followed again by placebos for two weeks. Throughout these seven weeks, tests were given to measure the criminals' aggressiveness and impulsiveness.

For the aggression test, each subject, one at a time, was seated in front of a computer screen as well as two buttons, labeled A and B. They were told that by pressing the button they were competing with the other subjects, who were located in a different building, and that they had the opportunity to earn money during the task. Pressing the A button one hundred times would earn the subject 15 cents. Pressing the B button ten times would take 15 cents away from another subject in the game.

While playing, the criminal would notice whenever money was subtracted from his total. This came across to him as an offense from another person. In reality, the subtractions were randomly generated by the computer and the subject was playing alone. Nevertheless, he was made to believe that a person took the money from him. In response, he could do one of two things. He could repeatedly hit the B button, supposedly in order to attack the illusory opponent. That is what the researchers considered the aggressive response. Alternatively, he could decide not to respond aggressively. He could simply keep pressing the A button to increase his own total. The researchers found that, while taking the antidepressants, the criminals chose the aggressive response much less often than they did while taking the placebos.

The test for impulsiveness had a similar design and similar results. It also involved buttons A and B. This time, however, clicking A twice earned the player 5 cents and clicking B twice earned him 15 cents. Here's the catch: following each click was a required wait time. The delay after clicking button B, the more profitable button, was much longer. Button A was considered the impulsive choice, because it granted quick gratification despite lower gain. Button B was the choice of self-control because the subject chose to delay gratification in favor of the greater good. The results of the study showed that the subjects also chose the impulsive response less often while taking the antidepressants.[37]

Some see these results as pointing to psychopharmacological cures for criminality. Antidepressants are one possibility, but there is also the option of antipsychotic drugs, which have been shown to pacify patients and prevent them from being violent. Those drugs, however, tend to work because they sedate the person rather than alter his judgment. Neither has yet proven to be the cure for criminal behavior. However, if new pharmaceutical methods are developed that can adjust the levels of all the relevant neurotransmitters in the body with more precision, we may actually be able to treat criminals before they commit their second offense.

But why must this intervention come after a person has already been labeled a criminal? If violent behavior is truly determined by the brain, as many neuroscientists suggest, we should be able to predict it before it happens.

This is the concept explored in the movie *Minority Report*, starring Tom Cruise. In it, police officers use visions from strange, prophetic beings to stop murders before they happen. Could police officers, alongside neuroscientists, do this today, using strings of neurobiological data? They might have the ability to establish whether someone is at high risk for committing murder, but they can't be *certain* of whether that person will actually murder someone. Nevertheless, the suggestion has been raised that the government should take action to contend with likely criminals before they commit a crime.

In 1996, in the United Kingdom, a man named Michael Stone murdered two women, a mother and her daughter, with a hammer. Some time later, it was revealed that Stone had been diagnosed as a psychopath. This led to a movement in Britain that encouraged the judicial system to take action against those who meet the biological requirements for psychopathy *before* they commit a crime. It was advocated that people in that kind of neurobiological state are at such high risk for violence that they should be dealt with preemptively. They are too dangerous to be kept on the street.[38]

In 1992, the first Bush administration introduced what it called the "Federal Violence Initiative." The project, led by Frederick Goodwin, then head of the Alcohol, Drug Abuse, and Mental Health Administration, was supposed to alleviate violence in certain areas of the United States by stopping it before it could happen. The plan was to identify youths who were supposedly at risk for criminal behavior and provide them with counseling or other measures to deter them from future illegal acts. Goodwin promoted the strategy by talking about the genetic and biological determinants of crime. Knowledge of the relevant biological data could be used to decide who is likely to become a criminal and give them the right education early. The idea was quickly shot down, accused of being a racist policy, since inner-city children were most likely to be targeted.[39]

The truth is that we do not yet have the technology to accomplish, with precision, any sort of predictive criminology. Of course, there remains the chance that such capabilities will exist in the future. Analysis of genetic, biochemical, and personal information about a person may one day reveal whether the individual is destined for greatness or imprisonment.

That may be the day when the appeal to neurobiological determinism will be a standard argument in the courtroom. If it is true that biological mechanisms are the sole cause of criminal behavior, then Clarence Darrow's defense of Leopold and Loeb should become universally accepted—not only to diminish responsibility for a crime, but to do away with it. If determinism is true, as the evidence we discussed suggests, then it must be that Leopold and Loeb were caused to murder by their brains. Their feelings of free will were illusions caused by the neuronal machines in their heads. They lacked the moral agency with which to inhibit the urge to do evil.

If Leopold and Loeb, brilliant young men with the most privileged of upbringings, lacked the sovereignty of will by which to countervail the impulse to murder, then perhaps this ability exists in no one. Should this conclusion ever be widely accepted, responsibility for immoral actions would become a thing of the past, eliminated in the face of the unquestionable, deterministic mechanisms of neurobiology.

As we have seen, the concept of determinism and the current knowledge of the brain leave open a disturbing possibility. It may be that, long before a violent criminal claims his first victim, his future evils are already determined, his life of immorality established. It may be that he is born to be a criminal. Conscious agency does not lead him down that path. His neurons make the decision to murder long before he first picks up the knife. How can he choose to live a moral life if the seeds of corruption are already planted in his brain?

13

MORALITY'S END

In the early years of the Cold War, it became clear to the aging Dwight Eisenhower that the United States was no longer the only country to have the atomic bomb. On everyone's mind was the very real threat of nuclear war. It was during this time that a new defensive strategy gained popularity among military and political leaders: that of "massive retaliation." The idea, proposed by Secretary of State John Foster Dulles, was to respond to all enemy acts of aggression with overwhelming force—even if that meant using atomic weapons. As Dulles put it, the country had to be prepared to immediately "strike back where it hurts, by means of our own choosing."[1] This way, the United States would avoid full-fledged warfare by intimidating the enemy early, giving any adversary a taste of America's devastating military muscle.

The doctrine of massive retaliation soon became the predominant military approach, but there were some who disagreed with it. The most notable opponent of the idea was Maxwell Taylor, who became the US Army chief of staff in 1955. Taylor didn't like the prospect of using nuclear weapons in future conflicts, or the concept of deterrence in general. He penned an article for the journal *Foreign Affairs* to make his objection public, but military censorship policies prevented him from publishing it. So Taylor was forced to swallow his reservations toward massive retaliation and focus on his duties.

By the end of his four years as army chief of staff, Taylor decided not to take a second term. Then, he was contacted by General Andrew Goodpaster, who brought a message directly from President Eisenhower. Goodpaster was aware that Taylor was not interested in continuing in his current post and came to suggest an alternative: he asked him to become the supreme allied commander of NATO, the first person to fill that role after Eisenhower. Taylor was stunned. The job was the most prestigious and honorable one he could imagine for himself, but should he take it? As far as his career was concerned, it was certainly in his interest. However, he disapproved of the military's way of dealing with enemy threats and, as long as he remained in uniform, his own opinions would have to be suppressed.

In order to uphold his personal beliefs and gain the ability to share them, at long last, with the public, Maxwell Taylor turned down the offer of a lifetime. He left the army completely and at once began work on a book on the problems of massive retaliation and the military's general approach toward warfare. He titled it *The Uncertain Trumpet*, inspired by a verse in 1 Corinthians: "For if the trumpet gives an uncertain sound, who shall prepare himself for battle?" The book moved many army officials to reconsider their views on massive retaliation and the use of nuclear weapons. It even had an influence on John F. Kennedy, who later made Taylor his personal advisor, a position that allowed Taylor to have a profound impact in the resolution of the Cuban Missile Crisis.[2]

It's safe to say that many people, especially those who read *The Uncertain Trumpet*, felt a deep admiration of Taylor for finding the strength of will to follow his moral convictions. Over the course of his service in the military, Taylor learned much about its operations. He gained a wealth of experience and eventually became one of the country's top military leaders. What his book shows, however, is that, throughout his career, Taylor reflected on his experience. As he observed the successes and failures of army missions, he contemplated the strategic elements that might have engendered the outcome. When he consulted with other officers, he listened to their opinions and, in his mind, deliberated about the points they made, asking himself how well they reflected the needs of the military and the country as a whole. From his years of careful thought and consideration, Taylor formulated his position against massive retaliation. And when the time came, his decision to refuse the offer of Eisenhower was based on principles that he had developed for himself through conscious introspection. It is for this reason that we admire him for his beliefs and respect him as a man of perseverance and moral courage.

There are some, however, who would contend that Taylor was not responsible for the development of his beliefs because they did not arise from any sort of moral agency. At no point during his military career did he have conscious, willful control over his developing moral convictions. Rather, all his beliefs were generated by his brain.

One proponent of such a view is neuroscientist Michael Gazzaniga. Gazzaniga has spent considerable time studying what he calls the brain's "left hemisphere interpreter." He believes that the activity in this region of the left hemisphere is responsible for producing beliefs—it receives information from all parts of the brain and generates an explanation to make sense of it.[3]

There is a phenomenon known as *split-brain syndrome* in which a person's corpus callosum (the thick cord that connects the right and left sides of the brain) is severed. This has been done intentionally, through surgery, as a treatment for epilepsy. The problem is that the procedure can have some strange effects on the person. Rather than one hemisphere being subordinate to the other, as in a normal brain, each individual hemisphere in a split brain might begin to work independently of the other. The patient may be left with what seem like two minds, *two selves*.

Gazzaniga says that studies of split-brain patients demonstrate the role of the left-hemisphere interpreter in producing belief. In one experiment, the word *walk* was shown to the right side of a person's brain. The patient stood up and began walking. When the patient was later asked why he began to walk, he provided a reason that Gazzaniga says was generated by the patient's left brain: "I wanted to go get a Coke."[4]

This patient's left hemisphere did not receive any information about the word presented to the right side, since the connections between the hemispheres were cut. Nevertheless, Gazzaniga says, the left brain used whatever information it had to yield an explanation.

Another example of how the left brain determines our beliefs comes from research on a cognitive disorder known as *anosognosia* (meaning "denial of illness"), which affects paralyzed patients.[5] Because of damage to the parietal cortex (the top surface region of the brain that extends from the middle area toward the rear), which might result from something like a stroke, these patients will simply *deny* that they are paralyzed. When asked about their paralysis, they tend to offer strange replies. Sometimes, patients will explain their inability to move a paralyzed limb by claiming that it doesn't belong to them, as shown by this doctor-patient dialog that was recorded by neurologist Vilayanur Ramachadran:

> Patient: Doctor, whose hand is this (pointing to her own left hand)?
> Doctor: Whose hand do you think it is?
> Patient: Well, it certainly isn't yours!
> Doctor: Then whose is it?
> Patient: It isn't mine either.
> Doctor: Whose hand do you think it is?
> Patient: It is my son's hand, doctor.[6]

How can this mysterious response be understood? Gazzaniga says that when the patient is asked about his or her paralyzed limb, the left-hemisphere interpreter has to cook up an explanation that reconciles two seemingly con-tradictory bits of information. The first is that, according to processing in the visual cortex, the arm is there and is properly attached to a body, but it isn't moving. The second piece of information is that, because of the damage in the parietal lobe (an area critical for movement), there is no available data about damage to the arm. How can the left hemisphere explain why a sup-posedly undamaged arm isn't moving? By generating the belief that it belongs to someone else.

An even more vivid illustration of how the brain generates belief is a disorder called *reduplicative paramenesia*. Patients with this syndrome will often have trouble conceiving of where they are in time, often mixing up the past and the present. These defective messages, Gazzaniga says, are sent to the left-hemisphere interpreter, which is working properly. The interpreter has to fabricate a story that reconciles those mistaken messages with what the person sees around him or her. Gazzaniga describes a patient with this disorder who came to see him at a hospital in New York City:

> This woman was intelligent; before the interview she was biding her time reading the *New York Times*. I started with the "So, where are you?" ques-tion. "I am in Freeport, Maine. I know you don't believe it. Dr. Posner

told me this morning that I was in Memorial Sloan-Kettering Hospital and that when the residents come on rounds to say that to them. Well, that is fine, but I know I am in my house on Main Street in Freeport, Maine!" I asked, "Well, if you are in Freeport and in your house, how come there are elevators outside the door here?" The grand lady peered at me and calmly responded, "Doctor, do you know how much it cost me to have those put in?"[7]

Gazzaniga says that this woman's belief that she is at home is produced by her left-hemisphere interpreter to accommodate her surroundings along with her mistaken sense of time.[8]

In his review of this research, Gazzaniga is saying that all of our personal beliefs are determined by the brain. The consequences of this view for our understanding of morality are no less than catastrophic. Morality as we know it assumes that we each have the ability to freely contemplate the moral nature of the challenges that confront us; to deliberate about the most ethical solutions based on our experience and moral framework, itself a product of years of conscious introspection; and will ourselves to take the course that best reflects our values and intuitions. If we are to suppose that our moral beliefs, the very foundations of our every ethical choice, are caused by the automated firing of neurons in the brain, then what is left to the agent? Perhaps the agent has no power over the content of beliefs, but rather only initiates a decision based on them. But how does the agent decide how to apply those beliefs in order to make the decision? The answer, no doubt, is that each person has another set of beliefs that dictate how to weigh our original set of beliefs when faced with a decision. However, *those too are beliefs*—and are therefore, according to Gazzaniga, caused by the operation of neurons over which we have no control. Suppose you decide that, in order to escape this deterministic prison, you will flip a coin or roll a die, or do the opposite of what your beliefs tell you. Is that decision free? No, because it resulted from a chain of causes and effects leading back to your feeling that you must escape the prison of determinism—a belief caused, yet again, by your brain.

The research done on violent criminals seems to suggest that the seeds of immoral behavior are planted in certain people by their brains, their genetic makeup, or their environment. The implications of Gazzaniga's view, however, are far more pervasive. His observations apply not only to immoral thought and action but to the totality of human behavior. Our beliefs—our reasons for acting as we do—are products of countless molecular and cellular processes, regulated by systematic chemical principles.

What, then, might we say of religion? If Gazzaniga's view applies to all human beliefs, then spiritual ideas must be determined by the brain, too. And there is evidence to suggest that they are. For example, there is a neurological condition known as *temporal lobe epilepsy* that has as one of its symptoms "hyperreligiosity," the propensity to become very religious and engrossed in moral thought. Gazzaniga says that temporal lobe epilepsy could be the reason why some people have powerful spiritual revelations. Vincent van Gogh, for example, had every symptom of temporal lobe epilepsy. The disorder may explain why he had many religious visions, such as one of Jesus Christ resurrected. Historical records suggest that, based on certain behaviors they exhibited, people like Muhammad, Moses, and Buddha may have had the disorder as well. Could that be why they became religious leaders?[9]

In a field of research that has been called *neurotheology*, neuroimaging studies of people engaged in religious acts show that both the frontal and the temporal lobes are involved. Scientists have actually attempted to induce spiritual experiences by exciting these areas of the brain. In one study, a magnetic field created by a helmet was used to stimulate specific areas of the temporal lobes of volunteers. As a result, the subjects reported having a variety of spiritual experiences. Some claimed to sense the presence of relatives who had passed away. Others described having out-of-body experiences. Still others felt contact with "another entity," possibly God or just some sentient being they did not recognize. These spiritual moments were caused just by triggering activity in their brains.[10]

Gazzaniga says that human beings are "belief-formation machines."[11] Our brains determine that we will have a certain set of opinions, values, moral principles, and, as we have just seen evidence for, religious experiences. The notion that our ideas arise from painstaking, conscious reflection on ourselves and our experience is sadly mistaken. Every belief that we have is generated by the biological mechanics of the brain, as a way of creating interpretations for what we see, hear, and feel, so that the human animal may be more adept at surviving in the natural world. It is those beliefs, determined as they are, that regulate our behavior. Our impression that our ethical decisions result from a powerful mental struggle to choose the moral path is, as Daniel Wegner would say, an illusion caused by our machinery. Thus *moral* and *immoral* are just terms that describe the interaction of the organic automatons known as human beings. They have nothing to do with deliberation or with willpower or with the inner conflict of good versus evil. All that is illusion. There is only the brain and nervous system. In the sense in which we understand it, *morality does not exist.*

The determinist attempts to redefine what we mean by morality and responsibility. As Michael Gazzaniga writes:

> Just as traffic is what happens when physically determined cars interact, responsibility is what happens when people interact. Personal responsibility is a public concept. It exists in a group, not in an individual Brains are determined; people (more than one human being) follow rules when they live together, and out of that interaction arises the concept of freedom of action.[12]

What does that really mean? It comes down to something like compatibilism, the idea that free will has nothing to do with conscious control of action, but with the availability of choices. As we established early on, compatibilism is simply determinism. It tries to save free will by saying that free will is something else—the ability to select from an array of options—but to the question of whether free, conscious deliberation controls our actions, the answer clearly is no.

Gazzaniga is attempting to make a similar last-minute save of free will. He says that we, as individual human beings, do not have free will. Maxwell Taylor does not have conscious sovereignty over his thoughts and actions. Free will arises from the interaction of people living together. Moral responsibility doesn't apply to any single person—it's just an idea associated with the way people interact. Moral agency is like traffic: it somehow comes about when determined people cross paths.

Of course, we know that the concept of moral agency is quite different from that. We have seen arguments showing that our beliefs are determined and evidence for a biological basis of spirituality. Gazzaniga is one of many neuroscientists who imply that such findings demonstrate the brain's deterministic supremacy over every living person. He is in the same camp as those who assert that the readiness potential disproves free will, those who believe conscious will to be illusory, and those who believe the effects of pharmaceutical drugs on the brain suggest a mechanistic, algorithmic nature for the mind. He shares a common scientific view that suggests that we, like Maxwell Taylor, are not moral agents but biological machines operated by the rules of neuronal interaction.

Gazzaniga is also one of many who go on to claim that ideas like freedom and responsibility can still exist, despite the brain's control of our decisions, because they come from the grouping of these determined persons. But this cannot be correct. To have free will is to have the conscious ability to deliberate over a question, to come to a decision, and to mentally command that one's body carry out that decision. If this capability is truly inaccessible to an individual human being, as Gazzaniga suggests, then it cannot be accessible to a group of human beings. To say that someone can be held responsible for an action is to say that the action can be attributed to a certain moral agent, since the action was freely and mentally initiated. But if the individual does not have free will—if it is true that the arguments for neurobiological determinism invalidate the possibility for freedom of action—then he must also be deprived of moral responsibility, whether he is alone or in the presence of others.

Gazzaniga's work is not an attempt to reconcile the feeling that we freely control our actions with the determinism implied by neurobiology. What we are seeing is the denial of free will, of responsibility, of moral agency, and of morality itself.

We should now come to realize that the denial of these essential human faculties is not an obscure viewpoint. The majority of modern scientists, if challenged with the right line of questioning, will concede that they cannot rationalize the existence of free will or the contingent doctrine of moral responsibility. In the face of overwhelming evidence for neurobiological determinism, these are concepts of which we simply must let go. Every human thought and action is reducible to the algorithms of neuronal processing. The mind is an antiquated notion. Free will is an illusion. Responsibility is a myth. It is meaningless for us to say that the moral agent now is gone, since what we really mean to say is that there never was a moral agent, only the illusion of one.

14

THE DEPTHS OF CONSCIOUSNESS

On December 8, 1995, French journalist Jean-Dominique Bauby, editor of the magazine *Elle*, experienced a terrible stroke and fell into a coma. Some time later, he awoke to find himself lying in a hospital bed. In moments, he noticed something else, something that meant that he could never return to his daily life: the stroke had left him paralyzed. Because of damage to his brainstem, Bauby, though fully conscious, lost the ability to move nearly every muscle in his body, a condition known as *locked-in syndrome*. Every limb, every inch of his torso, and almost every muscle in his face and neck were frozen. Bauby could do little more than blink his left eye. To any observer, Bauby appeared to be dead, a cadaver recumbent in a hospital bed.

But he was very much alive. Though his body lay still, his mind was surging with activity, a flood of pain and regret, longing and melancholy, but also of thought and reflection, ideas and creativity. Bauby was not content to equate his immobility with his extinction. He would not allow his mind to whither away. In his professional life, Bauby was a writer, and he decided that, despite his syndrome, he would keep writing.

Using only the blinking of his left eye, Bauby dictated to a friend, letter by letter, the words that he wanted written down.[1] He was shown a

143

page with a reorganized form of the alphabet, with the letters sorted by their frequency of use. His friend then pointed to each letter, one at a time, until Bauby blinked a signal indicating that the current letter was the one he wanted. Each word took minutes to write, so each one had to be carefully chosen ahead of time. Every day, before his friend arrived to receive his dictation, Bauby was writing paragraphs in his mind. Navigating his inner world of experience, he composed what he wanted to say. He sharpened his ideas. He thought about his word choice, adding adjectives and descriptive verbs. He erased, edited, revised, and refined. He did all of this in his mind, memorizing the final version, which he then dictated, letter by letter. After over two hundred thousand blinks, Bauby had completed a book, the title of which translates to English as *The Diving Bell and the Butterfly*.[2] Here is an excerpt:

> Through the frayed curtain at my window, a wan glow announces the break of day. My heels hurt, my head weighs a ton, and something like a giant invisible diving bell holds my whole body prisoner.... My diving bell becomes less oppressive, and my mind takes off like a butterfly. There is so much to do. You can wander off in space or in time, set out for Tierra del Fuego or for King Midas's Court. You can visit the woman you love, slide down beside her and stroke her still-sleeping face. You can build castles in Spain, steal the Golden Fleece, discover Atlantis, realize your childhood dreams and adult ambitions.[3]

The book discusses his experiences as a hospital patient with locked-in syndrome, as well as his thoughts on his past and life in general, his hopes and his dreams. It discusses the fact that, though his body is motionless, his mind is brimming with vitality. Bauby died of heart failure two days after the book was published.

Contemplating this remarkable story, one question comes to mind: is this, too, determined? Could Bauby's achievement really have been determined by neuronal calculations beyond his control? Something about that conclusion just doesn't sit well with me.

Let's think about this for a moment. Do we feel comfortable claiming that Bauby's conquering of his illness, his creative accomplishment despite the most devastating of obstacles, was determined by a set of formulas in the brain just as they might govern the motion of a stone down a hill? Had all the physical states of Bauby's brain been known to us in December 1995, could we have predicted that he would write *The Diving Bell and the Butterfly* by blinking his left eye? Or, could we have written the book for him, word for word, just because we understood the states of his brain? If we agree that Bauby's thoughts and actions were determined, these are the implications that we have to accept.

I'm sure you can agree with me that there is something faintly unsettling about this conclusion. The question we need to ask ourselves, however, is why we have this sense of discomfort. Is it simply because we are reluctant to accept determinism and give up our attachment to free will, or is there another reason as well?

Though I admit my general reluctance to discard free will and moral responsibility, I believe there *is* another reason why the determinist's conclusion about Jean Bauby's accomplishment is unpalatable. It has to do with the kind of evidence that exists in support of neurobiological determinism.

Let us consider that evidence for a moment. We began by discussing instances of brain damage, specifically those that supposedly lead to the impairment of free will (or whatever you want to call it). Recall that there are some who conclude that, since impairments to the brain cause impairments of the will, the will must actually be a determined brain process, and it must have been determined that Bauby would write a book about his experiences with locked-in syndrome. However, it should now be clear that this is a weak argument. Although certain types of brain damage can destroy our ability to consciously control our actions, that fact does not prove that healthy people lack free will.

Since our understanding of brain injury and its effects on behavior was insufficient to teach us about the nature of the will, we turned to specific neurological studies. The first was Damasio's investigation of somatic markers. Damasio theorizes that whenever we have an experience, especially one that impacts us emotionally, a kind of biological record is left behind in the form of somatic markers. The somatic markers that are generated can then influence our decisions without our knowledge. Damasio tested this hypothesis by evaluating the performance of patients with frontal lobe damage using behavioral tests like the gambling task, which requires subjects to draw from one of two decks, a risky deck with short-term gain and a safe one with long-term gain.

We went on to discuss the work of Benjamin Libet, who used electrodes to record neuronal activity in the brains of subjects while they flexed their wrists. Before the experiment, the subjects were instructed to record the time at which they experienced the conscious desire to flex their wrists. By demonstrating that this desire occurs after the appearance of the readiness potential (the spike in brain activity said to indicate the initiation of a voluntary action), Libet was able to conclude that our conscious desires do not cause our actions.

Daniel Wegner carried out a number of experiments to show that we can be tricked into thinking that we are using our free will when we actually are not. In one study, for example, subjects were tricked into thinking that their use of a computer mouse was causing a cursor to point to objects on a computer screen, but the cursor was actually being controlled by someone else.

More evidence for neurobiological determinism came from studies of behavior prediction. The principles of neuronal group coding studied by Apostolos Georgopoulos can be used to predict the general direction in which a subject's arm will move. From this we learned that scientists can use neuroimaging to figure out how our muscles will move before they actually do. Neuroscientists at the Washington University School of Medicine in St. Louis were able to predict whether subjects would succeed at a simple computer game (in which the goal was to determine the direction of a moving field of dots) by monitoring their brain activity. From these and other simple examples of behavior prediction, it has been concluded that *all* human behavior is, in theory, predictable and determined by the same kinds of neuronal processes.

We saw how the use of pharmaceutical drugs can drastically alter someone's personality and how changes in the levels of neurotransmitters in the brain can lead to criminal behavior. Finally, we saw Gazzaniga's claim that our beliefs are determined by the brain, that we are belief-formation machines.

I should take this opportunity to mention that the comments I will make on these experiments are simply my opinion. There are many scientists and philosophers who will be inclined to disagree. In fact, it is my understanding that, among neuroscientists, my view—one that defends free will and moral agency without compatibilism—is by far the minority opinion. The idea of free will is typically perceived as "unscientific."

Determinism is seen as a more elegant theory, one that seems more consistent with current scientific knowledge. The ability to wield moral agency, what I believe to be the quintessential application of human consciousness, tends to be shrugged off, thrown in with some compatibilist theory. There are plenty of people who will be ready to disagree with me. But, as philosopher Jerry Fodor has said: "If it isn't literally true that my wanting is causally responsible for my reaching... and my believing is causally responsible for my saying... then practically everything I believe about anything is false and it's the end of the world."[4] This is something I believe very strongly, and it is what I am trying to defend in this book. So starting now, I will be sharing more and more of my thoughts about the questions we have been grappling with. Let's begin by returning to our discussion of the scientific support for determinism.

I said that the discomfort we feel when we consider the deterministic understanding of Bauby's achievement can be, at least in part, explained by the kind of evidence used to argue for determinism. We might loosely divide that evidence into two categories. Suppose that the first category contains all demonstrations that involve damage to, or abnormality in, the brain. Let the second category include all studies done on healthy individuals.

Our discussion of the ways that brain damage impairs the control of action would fall into the first category. The same goes for our discussions of how pharmaceutical drugs and changing neurotransmitter levels affect human behavior. All these are examples of how changes to the brain cause changes in consciousness and the control of action. Together, they make the claim that, since changes in the brain cause changes in behavior, all human behavior must be determined by the brain. That claim might be expressed as follows:

1. If we have free will, then changes in our brains should not cause changes in our behavior.
2. Changes in the brain cause changes in behavior.
3. Therefore, we do not have free will.

Not only do these examples make the same case, but they also share the same weakness. This weakness is in the first premise. It is not reasonable to conclude that we have no free will only from the simple fact that changes in the brain cause changes in our behavior. We would all agree, for instance, that an F-16 fighter jet is controlled by a pilot. Assuming the fighter is not on autopilot, the person in the cockpit is in charge of the plane's flight pattern and weapons. However, someone might be inclined to deny this. He might claim that it is not the pilot who is in control, but the engine and weapons system. Why? Because if you were to damage the engine, by hitting it with a rocket, the plane's flight pattern would change. The plane might fly erratically or not at all. Similarly, if you were to short-circuit the weapons system, the plane might suddenly release its cache of bombs and fire its missiles, or it could be left defenseless, unable to fire at all.

It could be said that this implies that the pilot does not really control the plane—the engine and weapons system do. The argument might be written this way:

1. If the pilot controls the F-16, then changes to its engine and weapons system should not cause changes to its functionality.
2. Changes to its engine and weapons system cause changes in its functionality.
3. Therefore, the pilot does not control the F-16.

In this form, the parallelism between the F-16 argument and the above free will argument is apparent, as is the mistake in the first premise. Nobody denies that damage to the engine of an F-16 will change the way the plane behaves. By the same token, nobody denies that damage to a person's brain will change the way he behaves. These statements are true regardless of whether pilots actually control F-16s and whether people have free will. Taking neurological drugs is like modifying the internal processors of an F-16. Having a low concentration of serotonin or dopamine, and being violent as a result, is like having problems with the weapons system of an F-16, and releasing explosives as a result. The mere fact that damage to the brain impairs free will does not mean that it never existed to start with.

Though much of the support for Damasio's somatic-marker hypothesis was derived from studies of patients with brain damage, his work doesn't really belong in the first category of arguments. That is because Damasio does not suggest that the effect of somatic markers *eliminates* free will. Rather, they restrict its use. Recall that Damasio believes that somatic markers narrow our list of choices before we decide, but the decision of how to act is still left to the conscious mind. Unlike the above arguments, this *is* consistent. Consider the F-16 once more. Since damage to the engine or weapons system limits or removes the pilot's control of the plane, we *can* conclude that the pilot had *limited* control all along.

The same can be said for human beings. Even if we do have free will, our control over our bodies is not complete. We are limited by the capacities of our bodies. Conscious will does not grant every person the ability to dunk a basketball or derive the theory of relativity. Though they themselves may not be mechanical, the moral agent and the pilot of an F-16 both *depend* on machinery and are therefore limited in power.[5] However, we cannot claim that, because brain damage alters or terminates our behavior, we must have no free will at all. That reasoning is invalid.

If the first category of argument against free will is as faulty as we have seen it to be, it is probably not the source of our uneasiness when it comes to deciding whether Jean Bauby was determined to write his book. At least, I don't think it is. I think the trouble resides with the second category of evidence: studies done with neurologically healthy subjects.

The relevant commonality in these studies, I think, is the kind of tests they use. For those who deny free will, these tests represent an argument that the whole of Bauby's conscious deliberation, his every thought and blink that he used to compose his memoir, was determined. The power of that argument rests on the reliability of the experimental tests, on their capacity to help us understand Bauby's behavior deterministically.

Libet's test monitored people's brain activity as they flicked their wrists. Wegner had his subjects point to objects on a screen with a mouse, though the cursor was actually controlled by someone else. The work of Georgopoulos made it possible to predict the general direction in which someone's arm or leg moved. Neuroscientists at the Washington University School of Medicine in St. Louis had their subjects press one of four keys on a keyboard to identify the direction of motion of a group of dots. Monitoring of their brain activity revealed whether they would be right or wrong.

With these tests in mind, we ask again: how does the nature of the experimental evidence against free will explain our discomfort with the claim that Bauby's behavior was determined? In the brief review of the evidence, that explanation may have become a bit clearer.

Consider the vastness of Bauby's conscious deliberation. While lying still in his hospital bed, Bauby composed a volume of his deepest emotions—his anguish, his hopes, and desires. He filled it with his opinions, his humor, his sarcasm, and his creative ideas. Lacking the ability to communicate verbally, or with sign language, or with any familiar form of interchange, Bauby was able to express every idea by merely blinking his eye. He made a complete language out of an action that before was vacant of meaning. With this language, using only the power of his conscious faculties, he created a unique work of literature, an emotional expression of his struggle as a conscious person trapped in a disabled body.

Flicking your wrist, pointing a cursor on a screen, moving your arm, stating the direction of the motion of dots—these are the kinds of human actions that have been studied scientifically. The determinist asks us to believe that, since these actions have been shown to be caused by neuronal interactions, Bauby's action must have been caused in the same way.[6] As far as causality is concerned, his behavior is no different from the behaviors tested in the experiments.

But doesn't it seem different? There is something about what Bauby did that separates his behavior from wrist flexions and judging the movement of dots. The decisions and behaviors tested in these experiments just don't seem to be comparable to Jean Bauby's decisions.[7] They are...different. This difference is what makes us uneasy when we are told that experimental evidence points to the fact that Bauby's incredible achievement was determined.

Writing a book is certainly far more complex than flicking your wrist or looking at dots. This, however, cannot be the entire explanation. Somehow it doesn't feel right. Is the difference between deciding to flick your wrist and composing an entire book in your mind (and blinking every letter) only that the latter is more complicated? It seems, to me at least, that the distinction is a little more profound.

In composing his book, Bauby engaged in self-reflective, conscious deliberation. His behavior required thoughtful introspection and meaningful interpretation of his experiences. Bauby had to pick out and connect elements from the vast inner world of his experiences in order to construct a creative narrative of his life.

This kind of thinking, which could be said to be the essence of human decision making, has not been studied scientifically. The fact that someone can predict the direction in which you will move you arm or trick you into believing that you control the movement of a cursor on a screen does not, by any means, imply that human decisions, thoughts, and behaviors are determined. It appears that the hard evidence against free will doesn't really address the depth of Bauby's contemplation. The decisions studied were those, like wrist flexions, that require little forethought. As a result, we feel uneasy presuming, solely on the basis of those studies, that Jean Bauby has no free will. That conclusion seems premature.

Yet among scientists, determinism is the predominant assumption about human action. But if it's true that support for the theory is inadequate, why is it so widely believed? The answer is that determinism today is less of a theory than it is a worldview. Since so many things in nature are determined—chemical reactions, cellular processes, the motion of projectiles—many scientists choose to have faith that everything else is determined. The exceptions, of course, are the interactions that occur at the quantum level. Quantum mechanical processes are not determined. They are random. Aside from that, scientists will say, all other natural phenomena—including human thought and behavior—must be determined, since the phenomena studied so far seem that way. That is the leap of faith taken by determinists, though they believe the preponderance of evidence is on their side.

This leap of faith—the jump from saying that since physical interactions in the laboratory are determined, all human behavior must be determined—represents the greatest gap in their argument. Nevertheless, fortunately for the determinists, their worldview is particularly easy to uphold. If one is satisfied with that perspective, it takes no effort, after the fact, to point to the supposed causal determinants of any event. Consider the following, probably familiar, claims:

- The war in Iraq was determined because of George Bush's hatred of Saddam Hussein, who attempted to kill his father, caused him to initiate a war in his pursuit of vengeance.
- People insult other people because they are painfully aware of their own weaknesses and vulnerabilities.
- World War II was determined to happen because the Treaty of Versailles left the Germans feeling resentful, causing them to support the rise of Nazism.
- The writing experience that Jean Bauby gained while working on *Elle* magazine along with the suffering he experienced because of

locked-in syndrome caused him to write *The Diving Bell and the But-terfly* by blinking.

After an event has occurred, it's easy to attribute deterministic causes to it, but does one actually *know* that explanation to be true? The answer is that, no matter how reasonable the causal explanation may seem, nobody knows for sure. Though one event may precede another, it doesn't neces-sarily cause it. What's more, there is a difference between something being a *factor* or *influence* in bringing about an event as opposed to a *cause*. It may be true that a person's feelings of vulnerability may play a role in her deci-sion to insult others, but we cannot assume that this is *the* cause, or how often it even has an effect. Do we truly know that Bush had a desire for vengeance against Saddam Hussein and, if he did, that it caused him to support the Iraq war, as opposed to just influencing him? Was World War II really *determined* by the resentment felt by the Germans? Couldn't the resentment have been present and the war *not happened*? Nobody would have ever expected a patient with locked-in syndrome to write a book. Can we reasonably say that Bauby was determined to do so? Without hard evi-dence, these are all just hypotheses and assumptions.

So why are all these scientists so convinced? As I said, determinism has become a worldview, a lens through which people perceive the world. Determinists tend to view every event as being part of one long, causal his-tory—they interpret every event deterministically. When people interpret everything in such a way as to support their theory, it's remarkably difficult to convince them that they're wrong.

This effect can be seen from the following true story. In the 1970s, Stanford psychologist David Rosenhan decided that he wanted to learn how patients are treated in mental institutions. So Rosenhan and his col-leagues (eight people in total) contacted a number of mental hospitals claiming that they each heard voices in their heads. Once admitted, how-ever, the pseudopatients (as Rosenhan called them) acted normally, as they agreed to do before beginning the study. None of them described fake symptoms or acted strangely in any way. They acted as their everyday selves and responded to every question honestly. When asked about the voices in their heads, the pseudopatients said the voices went away.

Despite the very sane behavior of Rosenhan's group, the hospital staff interpreted whatever they did as an indication of some mental illness.

When asked about his family, one pseudopatient said that when he was a child, he was closer to his mother, but as a teenager he grew closer to his father and gradually more distant from his mother. What might seem to us to be a few perfectly normal changes in child-parent relationships was marked down by the hospital worker as follows: "[Patient] manifests a long history of considerable ambivalence in close relationships, which begins in early childhood.... Affective stability is absent."

Each member of the research group took notes on his or her observations. When nurses noticed this taking place, they marked their charts to indicate their findings of a new pathological syndrome: "Patient engages in writing behavior."

The members of Rosenhan's group stayed in the mental hospitals for between seven and fifty-two days. Each was released with a diagnosis of schizophrenia or "schizophrenia in remission." The fact that they were all sane was never detected. The hospital staff began by assuming that the pseudopatients were crazy and therefore, consciously or unconsciously, construed whatever they did in a way that backed up that assumption. It was only an actual mental patient who noticed: "You're not crazy. You're a journalist, or a professor. You're checking up on the hospital."[8]

The often dogmatic adherence to the assumption of determinism distracts people from a simple truth: the scientific, empirical evidence for determinism is vastly insufficient to be applied to cases of human decision making such as that of Jean Bauby. This is the source of our uneasiness when we see the attempt to deny his free will. The contrast between the decisions that have been studied and the depth of Bauby's conscious deliberation is overwhelming and cannot be ignored.

Free will is the commonsense view of human action. As we act, we feel that we are consciously, freely, controlling our behaviors. If that notion, which some might consider the most basic and essential to understanding our humanity, is to be done away with, we need to be given proof. The burden of proof, therefore, falls on the determinist.

In the meantime, the behaviors that scientists have tested and used to argue against free will and agency just seem different from decisions like Jean Bauby's. This distinction is something we have not yet fully explored. We hinted at it by discussing aspects of his decision such as creativity and self-reflection, but if it is true that Bauby is a moral agent, we must be able to say something more concrete. All we have shown so far is that there is some reason to doubt that all human thoughts and actions are determined, but what reasons do we have to believe in free will? This is the question left to consider, keeping in mind that our understanding of moral agency hangs in the balance.

15

A CHALLENGE FOR EXPERIENCE

The CIA station chief in Indonesia has been searching for ways to acquire intelligence about a terrorist organization known as Jemaah Islamiyah, a terrorist group responsible for murdering forty-five people, some of them Americans, when it detonated a bomb in a Hilton hotel in Jakarta. It also attacked the US Embassy in Bangkok and contributed to numerous attacks that have harmed the interests of at least six countries. The station chief believes they are planning further bombings.

On a Sunday night, the station chief receives a classified file saying that an opportunity has arisen. Agents in Jakarta have recruited a new member of Jemaah Islamiyah to spy on the group's activities and report back to the CIA. The recruit, code-named Accordion, was told that he must try to embed himself as deeply as possible into the terrorist organization until he reaches the inner planning circle.[1] That is the only way he would be given access to the group's crucial secrets.

After several months, Accordion sends word that he still has acquired no secret information. He has been given training in the building and planting of explosives but has not been told anything of interest. The leaders of Jemaah Islamiyah don't trust him with their secret plans and will not do so until he has proven himself to them. In terrorist groups, members have to demonstrate their loyalty and dedication to the cause by personally committing a terrorist act.

In Accordion's case, Jemaah Islamiyah has demanded that he orchestrate a car bombing. The terrorist group wants to kill an Indonesian police officer who arrested two of its members. In their fury, the terrorist leaders demand that Accordion prove his allegiance by building a bomb and planting it under the police officer's car. If he successfully carries out the assassination, he may become accepted by the group and given greater access to key intelligence that he can pass on to the CIA. If he fails or refuses, he will be cast out of the organization, or more likely killed, and the CIA will lose a crucial opportunity to stop the terrorist group's murderous activities.[2]

The CIA station chief has the final say in this matter and must submit his decision by Monday morning, giving him little time to find out more about the case. He must do the best he can with the information available to him. Is it morally acceptable for him to authorize Accordion to murder the police officer?

The station chief begins to consider the case. He wants Accordion to make his way into the inner planning circle of Jemaah Islamiyah and believes that his planting of a bomb will help reach that end. But how sure is he of that? It is possible that the terrorist leaders have no plans to reveal any information to him at all. Perhaps they already know that he is a spy and intend to feed him false information to mislead the CIA. Can he allow the police officer to be murdered when it is possible that no benefit could come of it?

Then again, he has a sense that these terrorists don't suspect anything. Accordion joined Jemaah Islamiyah of his own volition. The CIA recruited him only afterward. That means that there would have been nothing detectable in his behavior that would have aroused suspicion. But the question remains: will Accordion's bombing of the police car grant him enough trustworthiness among the terrorists to allow his acquisition of intelligence? What if the terrorists order him to commit another murder the following day? A dozen acts? How long can the CIA be expected to sponsor these acts of terrorism?

Drawing from his experience, the station chief believes this is probably not the case. The terrorist group just lost two of its members when the police officer arrested them. Their organization has probably fallen into some level of disarray. They are probably under more pressure to recruit members than they usually would be. What's more, their focus now is on vengeance. They want to kill the police officer, and do so as a grave warning to local authorities. Thus, the terrorists are distracted by local problems and probably not as cautious as they should be about spies from America. Accordion might gain their trust and their secrets, after all.

If the CIA allows the killing to happen, however, it will also be allowing the terrorists to send their warning to the Indonesian police. The murder of the officer by Jemaah Islamiyah could instill the police force with a newfound fear of the terrorist group. Officers might be dissuaded from going after other members of the group and the department might start to appease Jemaah Islamiyah.

There is also the issue of the way that this act will look to the American public. Of course, the CIA would try to keep it secret, but how long will that last? The information is bound to reappear sometime in the future, and the cost of this would be tremendous. It would damage the reputation of the American government with its people. The *New York Times* front-page headline might read "Investigators Accuse CIA of Approving Plot to Murder Policeman." Columnists would start talking about America's descent into lawlessness by using terrorism as a means of fighting it. People might wonder how bad the situation in America must be that we have sunk to such depths. The United States could lose the moral high ground in its foreign relations. Authorizing the murder could be crossing a boundary that we should not cross.

On the other hand, wouldn't the station chief be causing the death of an innocent man either way? If he prohibits Accordion from following the orders of the terrorists, it is almost certain that he will be killed. His refusal to carry out the bombing will likely be perceived as a sign of disloyalty. The terrorist leaders might make an example of him by killing him in front of the group. So if the station chief does not authorize the murder of the police officer, he will simply be allowing the murder of Accordion instead.

Actually, the killing of Accordion is probably not equivalent to the killing of the police officer. Accordion joined a terrorist group. Had he not been recruited by the CIA, he could have become one of the most dangerous members of Jemaah Islamiyah. Then again, he has not yet done anything wrong. The CIA cannot say that he deserves death when he has never committed a crime. The fact that he was willing to work with the

CIA also shows that he may have motives other than terrorism. Perhaps there *is* a moral difficulty with allowing him to be killed.

Still, it does not seem that allowing the killing of Accordion is as bad as authorizing the killing of the officer. For one thing, Accordion agreed to accept the mission, certainly aware of the risks. For all intents and purposes, he is working as a CIA operative, and thus may not be considered a civilian. He is a spy on a dangerous mission. Second of all, there is a clear moral difference between putting an agent in a dangerous situation that leads to his death and *directly sanctioning* the murder of an innocent man. Military commanders make strategic mistakes all the time. They send soldiers through minefields and into ambushes, and afterward they must carry some burden of responsibility for those soldiers' deaths. However, the moral problem is far greater for someone who directly sanctions—gives full permission for—the murder of an innocent man. *That* is the act that would damage America's moral reputation. That might do more damage to the United States than Jemaah Islamiyah ever could. That is why, the station chief decides, he cannot authorize the bombing of the police officer's car. He sends a message Monday morning indicating his disapproval.

Suppose that a scientist wanted to create an algorithm, a strict mathematical procedure consisting of a set of formulas, to engage in the moral reasoning necessary to provide an answer to the dilemma of the station chief. How would it work? The algorithm would contain a list of variables that represent moral principles, such as one against lying and one approving the protection of human life. To these and all the other rules, numerical values would be assigned. The more significant the principle, the greater the value. For example, the rule approving the protection of human life has a greater value than the rule against lying. Thus, if the algorithm were used to calculate a response to the question "Is it morally acceptable for someone to lie in order to save a life?" it would respond that it is acceptable. Formulas would compare the value of the rule against lying to the value of the rule for protecting human life and return a result in favor of saving life. For decisions involving several moral rules on each side, the algorithm would include steps to add the values of them or apply some formula to yield a result about which side is better. For a question like that of the station chief, the algorithm would contain steps that analyze huge bodies of statistics and calculate probabilities and generate predictions. Using extraordinarily complex formulas and calculations, the algorithm accepts the variables of the moral dilemma as input, manipulates them mathematically, and spits back the numerical result, which, according to the algorithm's designers, represents the proper moral course.

The way that the station chief contemplates and resolves the moral problem is vastly different from the way the algorithm generates an answer. They might both provide the same response to the dilemma, but the way they come up with the answer is completely dissimilar. The algorithm represents the dilemma as a set of rules, stored as arrays of numerical values, and then applies some mathematical function to generate the result. Yet there seems to be nothing formal, algorithmic, or mathematical about the way the station chief deliberates about his decision. He imagines the scenarios that could result from choosing one action or another, pictures them in his mind, reflects on them in the context of his experience, and uses his intuition and judgment to decide on the most ethical course of action. There is no formula or strict series of steps he is following.

Personally, I don't think it makes sense to represent the process of moral reasoning as a set of formal rules. You may have noticed earlier that the rules we listed to solve the dilemma did not really account for its complexity. One might be inclined to say that the reason for this is that we didn't list enough rules, but I would disagree with that claim. I think that no matter how many rules someone creates—no matter how complex the formulas or algorithms that are used—there is simply no way that a rule-based system can account for the depth of the human moral decision-making process.

Consider the decision of the CIA station chief. What kind of rules could we come up with that would fully express every aspect of the problem, his examination of it, and his strategy to solve it? On the surface, it may not seem like the station chief is dealing with that many concepts. There are a few relevant moral principles, such as the imperative to protect human life and the policy that it is better that one innocent person dies than hundreds die. There is the data on the dangers of Jemaah Islamiyah to consider, which is important in understanding the threat. There is also the information on how new members of terrorist groups gain the trust of their leaders.

Taken together, isn't this information enough to develop a solution to the problem? The answer is no. However many facts we may outline, we will not have enough to represent adequately the depth of the problem.

Why? Because there will always be an immeasurable number of details that we miss. Take, for example, the moral principle that one must protect human life. How do we understand what that means? Nobody ever gave us a formal proof—a list of rules or formulas—to defend such a principle. Our understanding comes from a wealth of knowledge about human history, about the interactions of people in the world. It comes from our grasp of concepts like death, suffering, weeping, struggle, growth, birth, youth, joy, hope, striving, and success—things we have learned about through experience. Our *experience* is what teaches us about the sanctity of human life. Not a list of facts or formulas.

When the station chief glances at the data regarding the history of terrorist acts committed by Jemaah Islamiyah or the statistics about how new members of terrorist groups move up the ranks, and makes judgments based on them, he takes for granted the vastness of experience that allows him to *interpret* the information he is given. To contemplate the significance of this data for his decision about Accordion, he has to understand what a terrorist is, what a bomb is, the rationale behind terrorism, the nature of violence, the emotional impact of terrorism, the power of fear, the idea of an enemy and the dangers of conflict, as well as the meaning of ranks, the keeping of secrets, the development of trust, what a promise is, loyalty, how agreements are made, how relationships develop, dedication, how people prove themselves, the strength of alliances, and countless other ideas. There are far too many concepts to list. I could fill the next hundred pages with others, but that still wouldn't be enough. Each idea on the list could have hundreds of pages written on it. How many books have been written on the notions of loyalty, promises, violence, human relationships, and the like? These are concepts that cannot be reduced to mere lists of premises or formulas. To understand them in any real sense, one must have experience.

The central role that conscious experience plays in human reasoning can be seen in the way we address a simple logic puzzle. Suppose that four cards have been placed on a tabletop in front of you. Each is labeled with a letter or number, as shown below:

Each card labeled with a letter has a number on the opposite side and each card labeled with a number has a letter on the opposite side. Now suppose that you are given the following rule: "If a card has a vowel on one side, it has an even number on the other." The puzzle is this: which cards would you have to flip over to test whether the rule is valid? Try it.[3]

The answer is that you would have to flip the first card (E) and the fourth card (7). Since E is a vowel, the first card had better have an even number on the other side; otherwise, the rule is false. The fourth card is labeled with an odd number, 7, and as such it had better *not* have a vowel on the other side. If it does, the rule is false. We don't care about the second card, because consonants have no place in the rule. We also don't care about the third card, the 4. Why? Because the rule only says a vowel on one side implies an even number on the other. It doesn't exclude the possibility that a card could have an even number on one side and a consonant on the other. So it doesn't matter what is on the other side of the middle two cards. We only need to flip the first and last ones to prove or disprove the rule. Did you get that? How long did it take?

When a group of university students was given this puzzle, the majority of the students got it wrong. However, when the exact same problem was posed in a slightly different form (to a different group), the vast majority got it right. Here is the new version: Four cards are lying flat on a tabletop. Each is labeled with the name of a beverage someone is drinking or the age of a certain person, as shown below:

You are given the following rule to consider: "If a person is drinking beer, he or she must be at least twenty-one years old." Which cards would you have to flip to prove or disprove the rule?

The puzzle is simple. The first card represents a person drinking beer. We had better flip that one to make sure she is twenty-one or older. The last card represents a sixteen-year-old. We had better flip that one as well, to ensure that he *isn't* drinking beer. As for the second card, anyone can drink Coke, so we won't flip that card. We also won't flip the third card, because there are no restrictions, in our rule, on what a twenty-two-year-old can drink. So the answer, as before, is that we should flip only the first and fourth cards.[4]

In this form, the puzzle is much easier. Most people tend to get it right. Yet they tend to have trouble with the original puzzle. The odd thing is that *they are both the same problem.* They have the same structure, require the same logical strategy, and have the same correct answer. A mathematical algorithm would follow precisely the same series of steps in each case, ignoring the differences in meaning, and solve them in the same amount of time. We, however, find the second form easier. We are better at getting it right and do so more quickly.

The explanation, I think, is that human reasoning is not purely rule based, or algorithmic. Like the CIA station chief, when we contemplate a problem, we appeal to our conscious experience, not merely to sets of formulas. In its original form, the puzzle is not strongly connected to any experiences we have had. It doesn't appeal to our picture of the world. It's just a logic problem that we have to work through mathematically. However, when the same problem is given in a way that appeals to our experience, to our understanding of the drinking age, for example, a light bulb turns on in our minds. We see from here a real-world example demonstrating that we appeal to our conscious experience when making decisions. Human reasoning is not algorithmic.

However we accomplish it, both we and, say, a computer running a mathematical algorithm can solve the logic puzzle. But only *we* can properly address the problem of Accordion and the terrorist organization. The reason can be derived from two things we already know. First, we know that a rule-based system, an algorithmic system, can only solve problems that are strictly defined in terms of rules, such as the logic puzzle we just worked through. The second is that the decision of what to do about Accordion is a *boundless problem,* one that is indefinable as a strict set of rules. What's the conclusion? That a rule-bound system is unable to address that decision. The argument goes like this:

1. A rule-bound system can only address formally bounded problems.[5]
2. The scenario of Accordion is a boundless problem.
3. Therefore, a rule-bound system cannot address the scenario of Accordion.

Recall our discussion of Jean Bauby, the patient with locked-in syndrome. When we considered the possibility that Bauby's writing of his book, *The Diving Bell and the Butterfly*, was determined by neuronal interactions, we experienced a certain discomfort, but we couldn't say precisely what we found unsettling about it. Now we can. The answer is already in our hands: *deterministic systems are rule bound.* If Bauby's act was determined, then his book—every word of it—must have been the output of a series of algorithms and equations. This means that, had we been given enough data, we could have mathematically derived the words that would be written on every page. His life experience, his suffering, his desires and deepest thoughts—all were part of the equation that determined the words that, through blinking, he would utter from his hospital bed. This is that problematic implication of calling Bauby a determined system.

When someone says that an event was determined, he means that it occurred as part of a natural chain of events dictated by the laws of physics. According to the natural laws in the world, that event necessarily had to happen. When someone says that a system is deterministic, he means that every event within the system is determined in that way, by formal laws. That is how all deterministic systems operate. They are bounded by a strictly defined set of rules.

If deterministic systems are rule bound, and no rule-bound system can address the problem of Accordion, then it must be true that no *determined* system can address the problem. Follow the logic:

1. A rule-bound system can only address formally bounded problems.
2. A deterministic system is a rule-bound system.
3. Therefore, a deterministic system can only address formally bounded problems.
4. The scenario of Accordion is a boundless problem.
5. Therefore, a deterministic system cannot address the scenario of Accordion.

What we have shown here is that the dilemma that we contemplated at the beginning of the chapter is one that a deterministic system cannot handle. But we *can* handle it, and that says something about us as human beings. It says that *we cannot be deterministic systems.* Think of it this way:

1. Deterministic systems *cannot* address boundless problems, such as that of Accordion.
2. Human beings *can* address boundless problems, such as that of Accordion.
3. Therefore, human beings are not deterministic systems.

We can solve problems that determined systems cannot solve. Therefore, we ourselves must not be determined systems. That is what I contend, at least.[6]

Early on, when we first introduced the problem of neurobiological determinism and responsibility, we put it as follows. Suppose that at time A a person becomes aware of a problem and at time B he decides on a resolution. Assuming that there is little to no externally caused interference with his deliberation, if the states of his brain at time A determine his decision at time B, free will and moral responsibility do not exist.

In accepting the position that we are not deterministic systems, I am saying that the brain states at time A *do not* determine the decision at time B.

At time A, Jean Bauby woke up in a hospital bed to find himself helplessly immobilized by locked-in syndrome. At time B, Bauby completed the book *The Diving Bell and the Butterfly*, an inspiring memoir of mental life in an otherwise lifeless body. In the time from point A to point B, Bauby had little distraction from his conscious introspection. He thought about his task of completing the manuscript and his struggle against the decay of his conscious faculties.

At time A, Sunday evening, the CIA station chief receives the file on the possible Accordion mission. At time B, on Monday morning, the station chief indicates that the mission should be aborted on ethical grounds. As the decision is made overnight, the station chief cannot gather more information about the case than he already has. He can depend only on his experience and moral deliberation to resolve what he should do.

In my view, the brain states of Bauby and the station chief at time A could not have determined their decisions at time B, because that would imply that some mathematical algorithm generated the words in *The Diving Bell and the Butterfly* and the intricacies of the station chief's moral reasoning. I think that the challenges of composition in creative writing and moral consideration in strategic intelligence are boundless problems that cannot be addressed by solely rule-based systems. We, however, can address them and do so with such success that literature and records of past moral decisions are the primary tools we use to gain insight into human history and the evolution of society.

To show the distinction that I think exists between human reasoning and algorithmic processing, we don't need to use examples as intricate as the above two. The differences are apparent even in simple decisions, moral or otherwise. For example, consider the following incident.

On a Friday evening, a metro bus returns to its main terminal with a wide dent in its left side. There are no passengers on the bus at this late hour. When asked about the damage the next morning, the driver of the bus says that he was the victim of a hit and run. He says that, on his way to the terminal, he parked the bus and went for a quick walk to stretch his legs. When he came back he saw the dent on the bus's left side and noticed that whoever caused the damage did not leave a note.

The next day, it is reported that a tree has been damaged not far from the main bus terminal. Blue paint has also been found around the damaged region. Furthermore, an inspection of the bus reveals that there is an empty bottle of vodka under the driver's seat.[7]

Suppose that you are asked two questions about this case.

1. What happened to the bus?
2. Who is morally responsible for the accident?

To answer the first question, we might first want to ask a question of our own, namely, "What color is the bus?" As we might have guessed, the bus is blue. So with that in mind, it seems plausible that the bus ran into the tree, causing the dent in its left side, and the bus's blue paint rubbed off on the tree. This possibility, taken together with the finding of an empty bottle of vodka beneath the driver's seat, makes it seem likely to us that the driver was lying about the hit and run incident. The best explanation seems to be that the driver was drinking on that Friday night. With his judgment and reflexes impaired, he lost control of the bus on his way to the terminal and hit a tree. That is how the dent got there. That is also why the answer to the second question is obvious to us: the driver is morally responsible for the damage to the bus.

This resolution of the problem certainly isn't difficult to come up with.

The scenario is far simpler than that of the station chief. Yet, as before, our understanding of the problem and the strategy to solve it cannot be expressed as a set of rules. As in the Accordion situation, there are an immeasurable number of relevant concepts that we can take into account as we contemplate the situation. It is a boundless problem.

We have to know, for example, that buses are mobile vehicles that travel along roads, which are hard surfaces with the ability to support their weight. Buses have drivers who have the ability to control their direction and speed of motion. Drivers are generally supposed to keep the bus within a lane. The permissible lanes are on the right side of the road. However, drivers some-times make mistakes and cause the bus to swerve in one direction or another. We also have to understand that vodka is a popular drink, that vodka con-tains a high percentage of alcohol, that alcohol can impair reflexes, that impaired reflexes can lead a driver to lose control of a bus, that an uncon-trolled bus is susceptible to having an accident, that trees often grow along roads, that trees are hard, that poorly operated buses can strike trees, that buses are made from metal and can be dented by intense forces, that trees are stationary, that the impact between a moving bus and stationary tree can dent a bus, that paint can be transferred between objects by contact, that dents in metal do not spontaneously repair themselves, that a dent in the side of a bus is a sign of an accident, that accidents tend to be investigated, that investigators of vehicular accidents tend to seek the person who is respon-sible for them, and that an intoxicated bus driver can be held morally accountable for any accident that occurs while he is at the wheel.

And that is only a taste of what kind of experience it takes to resolve the problem. We need to comprehend much more about buses, roads, traffic, trees, metal, paint, alcohol, intoxication, judgment, reflexes, speed, collision, lying, and the motivations and behavior of bus drivers.

An algorithm could not analyze this scenario in the same way that we do. The amount of information that could be relevant to the situation is endless. There is no standard way to determine which information is rele-vant. And once relevant concepts are discovered, there are countless ways to interpret them. How could a strict series of steps be used to decide what subtle concepts are necessary and appropriately integrate them into the construction of an explanation? There is no intrinsic connection between the presence of a glass bottle in a bus and the fact that the bus is damaged. What's more, the rules of logic do not dictate that the discovery of such a bottle should indicate anything about the cause of property damage or the honesty of a public transit employee. It takes a conscious being to navigate his inner world of experience in order to make that connection and contemplate its significance and consequences.

The implications of this simple example are the same as those of the complex examples: since we can address boundless problems and determined systems cannot, we must not be determined systems. To put this in numbered form once more:

1. Deterministic systems cannot address boundless problems.
2. Human beings can address boundless problems.
3. Therefore, human beings are not deterministic systems.

This is certainly an important conclusion, but we have not yet established that we have free will or moral agency. It may certainly be the case that human reasoning is not determined—not rule based and not algorithmic—but that doesn't mean that we have free will. That does not mean that we are moral agents. Though we know that the human conscious decision-making process is not determined, possibly the greatest question remains unanswered: if not by a set of formal rules, *how else could the mind work?*

16

THE BOUNDLESSNESS
OF REASON

It is dark at the American military encampment when the storm begins. The rain falls in torrents, drenching the dozen or so soldiers who are out on patrol. The blackness of the sky is disturbed only by the sporadic flash of lightning. At his post, a half mile from the encampment, Staff Sergeant Charles Westfall is growing nervous. His squad of ten has been given the standard order to patrol and stand guard for the night, but the onset of the storm is making the situation increasingly difficult to manage. The relentless downpour has made it nearly impossible to see clearly and, worse, owing to some technical failure, all communications are down.

As part of a platoon of thirty soldiers, he and his men arrived here, in enemy territory, only two days ago. They chose to set up camp at this location because of its strategic proximity to an enemy base, which, to Westfall's knowledge, remains unaware of his company's presence. Fortunately, the only visitors to the camp have been roaming deer and lost civilians from a nearby town. The only problem with the site is that it is close to an area that may still contain old mines from a battle that took place years ago. The field was supposedly cleared, but military intelligence suspects that some of the mines were missed and are still buried in the soil. To be safe, the platoon has been instructed to stay clear of the area.

This recommendation provides Westfall with little comfort as he hears what sounds like a small explosion in the distance. His heart begins to pound. Either one of his men discharged a weapon by mistake or someone stepped on a mine and is critically injured out there, he thinks to himself. His eyes dart around in all directions. In this weather he can't see any of his men, who have spread themselves throughout the site. The sound came from the area ahead of him and to his left—around the suspected mine-field. Water streaming down his face, Westfall picks up his sniper rifle and slowly scans the field. What happened out there? Suddenly, through the sheets of rain he sees what appear to be human figures approaching from that direction. They are blurry on the scope but, from what he can make out, there are between three and five of them. They seem to be moving briskly toward the site of the explosion—and toward the encampment.

Who would be approaching the camp from that direction? A thought occurs to Westfall: could those men be enemy troops? They could have come from the base located only several miles away. The camp is not heavily guarded and Westfall's orders are to shoot at anyone who threatens it, though he thinks it would be wise to avoid giving away his squad's position unless it is absolutely necessary. However, Westfall is aware that his men have a great advantage over the enemy: their sniper rifles have significantly longer range than the enemy's machine guns. Westfall knows that if he is to engage foreign soldiers, the most effective strategy would be to do so immediately once they come in range, before they get too close. That way, the enemy will be unable to return fire because their bullets can't go far enough. If the approaching figures *are* enemy soldiers, now is the time to begin firing.

But what if the figures are members of his squad? What if they are of some neutral party—civilians, perhaps? Westfall is faced with a boundless problem, such as that of the CIA station officer, that of the investigator of the bus accident, and that of Jean Valjean. With his communications down and visibility limited, Westfall must use what information he has to piece together what has happened on the battlefield, as well as to decide how to respond. And time is running out.

Westfall begins to think of possible explanations for what he saw and heard. Drawing from his experience and the context of this situation, he constructs the following list in his mind:

1. **One of my men discharged a weapon by mistake.** The figures in the distance are members of my squad who are coming to see what happened. *Don't shoot.*

2. **An enemy soldier discharged a weapon.** One of my men may have already been killed by an enemy grenade or gunshot. The figures in the distance are guerrilla fighters who have found the encampment. They are beginning an attack. *Shoot!*

3. **One of my men stepped on a mine.** The storm has made navigation in the area difficult. A friendly soldier got off course and ended up where he shouldn't have. The figures in the distance are members of my squad who are coming to see what happened. *Don't shoot.*

4. **An enemy soldier stepped on a mine.** They probably don't know about the old minefield. The figures in the distance are guerrilla fighters who have found the encampment. They are beginning an attack. *Shoot!*

5. **A deer set off a mine.** This area is filled with wildlife. A deer or some other animal was running around in the dangerous part of the field and set off a mine. The figures in the distance are members of my squad who are coming to see what happened. *Don't shoot.*

6. **We have been betrayed.** One or more members of the platoon revealed our location to the enemy. They have come with enemy troops to attack the camp. One of the traitors or enemy soldiers stepped on the mine. The others are behind him and attempting to ambush my men on patrol. *Shoot!*

7. **A civilian stepped on a mine.** Someone from the nearby town was walking home with his friends. He set off a mine and his friends are running over to see what happened. *Don't shoot.*

8. **An enemy scout stepped on a mine.** The enemy does not know about the encampment. They are not coming to attack. Those figures out there are scouts on patrol who wandered into the old minefield. One of them happened to set off a mine. If they see the encampment, they will report its location to their superiors. *Shoot!*

9. **I shouldn't trust my senses.** I have been out here with my sniper rifle for hours. It's the middle of the night and I'm exhausted. I can barely see anything in the rain. I didn't hear an explosion—only the distant rumble of thunder. Those aren't people out there. Those are trees and bushes moving in the wind. *Don't shoot.*

So far, in his conscious deliberation, Westfall has construed the problem at hand, formulated a number of possible explanations for what occurred, and thought quickly about how he personally should respond (whether he should shoot). His decision to fire or not fire his sniper rifle depends on the explanation he accepts. So his mind moves on to contemplate each scenario he created, to find the one that is most plausible.

One possibility he eliminates easily. By simply pausing for a few moments and watching carefully, Westfall is certain that he is neither confused nor hallucinating. There are definitely figures moving toward him in the distance. But who are they and why are they there? His deliberation continues:

"Could it be that a deer set off a mine, and those figures are my men coming to check what happened? That is doubtful. They appeared only seconds after I heard the explosion, meaning that the explosion was in their range of sight. They would have seen that it was a deer. Then again, the storm severely limits their visibility. However, that doesn't change the fact that they were close enough to the explosion to distinguish the body of a deer from the body of a man. And they wouldn't risk approaching the minefield to check a deer carcass.

"A civilian couldn't have set it off either. Who would be out taking a stroll at this late hour, in pouring rain, several miles from the nearest town?"

After rejecting those possibilities, Westfall goes on to consider the ones that remain:

1. One of my men discharged a weapon by mistake.
2. An enemy soldier discharged a weapon.
3. One of my men stepped on a mine.
4. An enemy soldier stepped on a mine.
5. We have been betrayed.
6. An enemy scout stepped on a mine.

"The sound of an exploding mine is distinct from the sound of any handheld weapon that our platoon or the enemy carries. Even in this storm, despite the pounding of rain and the crack of thunder, I don't think I misheard the explosion. The sound was very much like that of an activated mine and very unlike that of a firearm. Nobody discharged a weapon, either mistakenly or intentionally.

"Someone stepped on a mine, then, but who? A soldier in my squad? A traitor? An enemy soldier? A scout? It couldn't have been someone in my squad. Everyone under my command has been warned about the minefield. None would be so foolish as to venture near that area, no matter how bad this storm is. Those figures also can't be my men because I sent them to patrol individually. There is no reason that explains how three or four of them ended up together. What about a traitor? Could he be leading enemy troops against the camp? No. Even if someone has betrayed me, and is bringing the enemy this way, he would be sure to lead them around the minefield.

"So it must be an enemy soldier or scout that set off the mine—either way it seems that I should shoot."

Westfall readies his weapon. As he does so, however, he begins to question his thinking. What if the figures are enemy scouts who do not know about the camp? His initial thought was to fire at them anyway, but he now realizes that this would not be the best course of action:

"If I shoot now, I will give away my position—the position of the encampment. If those men are scouts and even one of them escapes, he will return to his base and report our location to his commanders. The operation here would be completely sabotaged and by morning an entire enemy battalion could overwhelm us and kill everyone in the platoon. If those are truly enemy scouts who are unaware of the American presence, I should not give away our position by firing at them.

"However, it seems unlikely that enemy scouts came miles from base during standard patrols and, by coincidence, ended up here. No, whoever those men are down there, they know we're here. They are enemy soldiers. One of them stepped on a mine. The others will continue to approach until they are in range, then they will begin their attack."

Westfall accepts this scenario as most plausible, but he does not begin shooting. Though he knows that, if those truly are enemy combatants, he should take them out, he decides that there is too much room for doubt in the situation to warrant aggressive action at this point. He decides it would be irresponsible to begin shooting at unidentified persons before he can acquire more information about who they are and what their intentions might be.

In this story, Westfall is able to think through a boundless problem and decide on a resolution. We call the problem boundless because, when presented, it lacks form. It lacks any guidelines to help identify what information may or may not be relevant. More important, the dilemma cannot be expressed as a set of strict rules or formulas that would allow it to be solved by applying an algorithm. Yet Westfall manages to contemplate his situation, create a set of possible explanations, choose the most plausible one, and decide on what he believes is the best course of action. How does he do it?

Though it would be ideal to have a complete theory of conscious decision making to answer that question, neuroscience simply isn't advanced enough to provide one. At this point, nobody can reasonably create a precise model of how the mind works. However, I believe that I can say enough about it to demonstrate that conscious will and moral responsibility are plausible. I cannot definitively prove the idea I am about to lay out (in this debate, not much can be definitively proven until we learn more about the brain), but I can show that it is consistent with our sense of how decision making works as well as the neurobiological literature of decision making. I can find examples to support the theory from those areas and argue that my view is better than neurobiological determinism, which we have already shown to be seriously flawed.

The idea, which I will call *reflective introspection*, is this: we make decisions, in a way that is not algorithmic, by consciously navigating our inner world of experience and reflecting on or reinventing the ideas and connections we discover. In the midst of his deliberation, Westfall does just this.

The story begins as Westfall is presented with two barely intelligible sensory stimuli: a noise and a blurry image. Nothing about those perceptions immediately indicates that a decision needs to be made on his part, but he begins to consider that possibility anyway, intuiting that something is amiss. What's more, Westfall is able to create from these disconnected stimuli a structured problem in the context of which he can direct his inquiry. As he watches the distant figures approach his camp, he ascertains that he has been faced with a predicament that calls for him to decide how to confront a possible threat and that his decision will have strong moral consequences. Westfall didn't have to construe the problem that way. There is no rule in logic that necessitated that interpretation. He could have construed the situation many other ways. For example, he could have interpreted it as a social problem. In general, when people are approaching your place of residence, they are visitors. Because of the poor visibility and broken lines of communication, they might have fired off a blank to signal their arrival. Westfall could have begun worrying about the cleanliness of his camp. Has food been prepared for these guests? Have beds been set up for them to sleep in? As a staff sergeant for the US military, what ought he do to ensure that the visitors have a pleasant stay and that his men do not come across as rude or inhospitable?

He could have also construed the situation as an athletic exhibition. Perhaps the noise was made by a gunshot, indicating the start of a race. The men in the distance are runners competing in the 1,500-meter event. The problem in Westfall's mind could have been how to find a better seat to watch the race. Why should he, an officer in the military, have to squint to watch a sporting event? After his years of dedicated service, does he not merit a front-row seat? Would his superiors castigate him for leaving his post to get a better view?

By appealing to his experience, Westfall is able to construct a difficult, consequential problem for himself out of disconnected stimuli from his environment. He takes that problem from the abstract to the practical as he wonders whether to fire his weapon.

Westfall realizes that his decision to either shoot or not shoot depends on what the sound he heard and the blurry image he sees represent. There are, however, an infinite number of possible explanations. Lacking any guidelines to help him, Westfall searches within himself for possible scenarios that meaningfully connect the information with which he has been presented.

Navigating his inner world of experience, Westfall picks out concepts and memories that he believes to be relevant. He takes into account the placement of his men in the field, recalling the orders he gave them. He considers the proximity of the enemy base and of the civilian town as well as the likely tendencies of military scouts and civilians to roam this area at night in bad weather. He decides that it is worth his time, despite his limited window for action, to think about the number of deer he has seen come out of the woods lately, since deer would not know to avoid the minefield. He even considers the possibility that his perceptions are unreliable, aware that they are especially fallible during a storm like this one.

From the seemingly limitless amount of information to which he has access, Westfall chooses to reflect only on the elements he finds most significant, leaving alone those that he finds insignificant. For example, he does not factor into his decision the price of the bullets in his rifle or the risk of wasting them. Westfall knows exactly how much they cost and that bullets should not be wasted, but his understanding of the situation tells him that those details are not significant in his current context and need not be weighed into his deliberation. Wasting bullets is the least of his concerns. Then again, had his men been very short on ammunition, Westfall might have taken the availability of bullets into account. If he was aware that, in the entire platoon, there were only ten bullets left, he would become wary of using one of them unnecessarily, and thus more reluctant to fire on the approaching figures.

Westfall could have considered matters of etiquette, engrained in him by his parents, in deciding whether to fire his rifle. He knows that most of the platoon is asleep in their quarters and that it is impolite to wake those who are sleeping. He also knows that he is near enough to the encampment that the gunshot would be heard. Yet by navigating the halls of his mind he is able to decide that the rules of etiquette do not apply in this context.

Beyond both his construal of the problem and his exploration of possible scenarios and responses, Westfall subjects his dilemma to yet another, higher level of analysis: he is able to reflect on the mode of his deliberation. He wonders whether he has thought about his options correctly. Westfall recalls that he has made several mistakes over the past few months for which he has been reprimanded. Perhaps a fear of committing another error has clouded his judgment. Is he being overly cautious? Was his decision made too hastily? It could be that he needs more time to come to a full conclusion and that he must reserve judgment. Are his intentions pure? Is he factoring his own emotions too heavily? At every stage of his deliberation Westfall can question himself and reflect on his purposes and modes of contemplation. He can reflect on the way he wanders in his mind. This is the essence of reflective introspection, and it is this capacity, I believe, that constitutes human free will.

Reflective introspection is what leads Westfall to refrain from pulling the trigger. It is the intentional control over the mind that he uses to resolve his dilemma. However, the story began with an algorithm over which Westfall had no control: the algorithm of perception. Westfall had no control over the raw information processing of his visual and auditory systems. Yet their activity, which allowed him to hear the explosion and see the figures, led to the start of his deliberation and thus represents the first step in Westfall's decision.

The first step of his decision was algorithmic. The last step of his decision was introspective, and therefore nonalgorithmic. What does that mean? It means that somewhere along the line there must have been a shift from algorithm to introspection. If it is true that such a transition takes place—that, though beginning with algorithms, our choices are resolved by reflective introspection—then we have built ourselves a foundation for defending the plausibility of free will and moral agency. In what follows, I would like to characterize this transition and show that it is consistent with the way human reasoning seems to work, and, finally, argue that this consistency is evidence that reflective introspection is truly a plausible account of how we deliberate.

In his somatic-marker hypothesis, Antonio Damasio theorizes that whenever we have a conscious experience, biological markers are left behind. Those markers represent connections that were made between the stimuli we perceived during the event and how we felt about them. The somatic-marker system is one that links the algorithmic processes of perception to the conscious reflection of the mind.

The brain is constantly processing data. Streams of information flow through the cerebral cortex from countless sources. Somehow, all of it gets organized. The results of visual processing in Westfall's brain are connected with the results of the auditory pathway in order to synchronize his experience of seeing rain with his hearing it. Such connections between separate pieces of data are made every second, and not only between sensory data.

Sometimes, as Damasio says, neuronal processing will detect a connection between some fragment of incoming data and some emotion, memory, or conscious experience we have or have had. This could occur with the help of a system much like that of the somatic markers, one probably centered in the frontal lobe. The significance of this connection cannot be established by any algorithm—*there is no formula for meaning.* So how is this connection, made by an algorithm, supposed to aid in the success of the human being if that algorithm can't interpret it? Does it just go to waste? No, it is transferred to the system of the brain that can reflect on its importance: the conscious agent.

Westfall's decision to shoot or not to shoot begins with algorithms. These determined, neuronal processes are at work constantly, keeping his heart beating, his lungs contracting, and processing all the data of perception. They are what allow him to feel the rain streaming down his face, hear the explosion on the field, and see the figures approaching in the distance. So far, the conscious agent is quiet, merely shifting attention from place to place in search of new and interesting information. However, unbeknownst to Westfall, a thought is about to strike him, one that will call upon the power of his conscious faculties.

Deep within his frontal lobe, algorithmic processes have identified a link between a certain auditory stimulus and a set of conscious experiences from his past, stored as memories. This connection being quite strong, it is sent to Westfall's consciousness for further inspection.

What does Westfall experience? A feeling. An instant of fear and anxiety. The sense of danger. Moments later, another link is detected, one between a number of visual stimuli, the previous auditory stimulus, and another set of conscious experiences. Westfall is struck with the feeling that something is wrong. Something is out of place. There may be danger.

More than this, algorithm alone cannot tell him. What has happened? How should he act in response? The algorithms that sent the data to his mind cannot even ask those questions. How could they? They are just mathematically processing information. All they can do to initiate further interpretation of the data is pass it on to the nonalgorithmic, conscious component of the brain—the moral agent.

But that is enough. Once presented with the connection, Westfall's conscious mind achieves an understanding that an explosion has occurred and that several unidentified figures in the vicinity of that explosion—an area he knows to be a minefield—are approaching the camp. Westfall then construes the situation as a volitional problem. What should he do? Should he shoot at them? Well, that depends on who they are and what they want. With this realization, Westfall begins to formulate a number of explanations: "Perhaps one of my men went off course and stepped on a mine. Alternatively, those could be enemy soldiers approaching, one of whom set off the mine. Maybe it was just a deer. Perhaps I should not trust my senses. I can barely hear or see anything out here."

Now, more immersed in his deliberation, Westfall begins to pass judgment on each scenario he developed, seeking the one that is most plausible. Upon finding it, he decides that though the figures are most likely enemy soldiers, it would not be prudent to fire on them now. The power of his conscious will keeps his gun barrel quiet and inactive.

What begins as an algorithmic data connection transitions into a problem for the moral agent. It evolves in his mind until Westfall resolves what action he should take and wills his body accordingly. The shift of control is the key, and that could happen by a mechanism much like that of the somatic-marker system—one that bridges the algorithm of perception with the self-reflective capacity of agency.

What happens when the bridge fails? What happens when the linked information cannot be sent to the conscious agent? It cannot be expounded upon. It remains within the algorithmic system. Recall Damasio's gambling experiment, in which he sat a subject in front of four decks of cards. The cards could either earn the subject money or take some away. The amount earned or lost depended on the deck. Decks A and B (the impulsive choices) yielded more money initially, but caused greater loss in the long run. In contrast, decks C and D (the prudent choices) yielded money gradually, but were the better choice for the subject because they gave him or her the most money in the end. When healthy subjects were given the task, they eventually caught on to the fact that picking from decks C and D was most beneficial to them. However, patients with damage to their frontal lobe never really discovered the best strategy. They tended to stick with decks A and B for the short-term gain, even though that tactic was hurting them. Damasio concluded that the damage to their somatic-marker system in the orbitofrontal cortex (a region in the frontal lobe) compromised their ability to make decisions.

I am going to add to Damasio's explanation. When the lesions to the patients' frontal lobe damaged the somatic-marker system, they undermined the transition between algorithm and introspection. The patients' ability to clearly reflect on their situation and decide on a more advantageous plan of action was impaired. Algorithms contributed more to their actions than agency. As a result, their behavior was more impulsive, less thought out—more focused on instant gratification than long-term goals.

This is the type of behavior often thought of as being *animalistic*. It is very likely that the more primitive animal species are not self-reflective. They are not moral agents. Rather, they are determined, mechanical systems—patterns of organic matter that operate solely through algorithm. They are machines.

Our ability to go beyond the algorithms—to achieve understanding, to appreciate meaning, to imagine, to consciously deliberate, to reason through boundless problems, and to act as free agents—is what separates us from the lower animals, from computers, and from all other machines. It is what makes us human.

The mind endows us with the ability to reflect on the connections made by lower-level, algorithmic processes. This is what allowed Westfall to find significance in the fact that he has seen deer in the past few days, but not in the fact that his ammunition should not be wasted. This ability selectively to expound on the ideas we discover while exploring them in our minds is essential, not only when we make decisions, but also when we have conversations, when we need to clarify situations that confuse us, when we imagine, and when we consider artistic creations. Consider, for instance, the way a student interprets Robert Frost's famous poem "Stopping by Woods on a Snowy Evening":[1]

> Whose woods these are I think I know.
> His house is in the village, though;
> He will not see me stopping here
> To watch his woods fill up with snow.
>
> My little horse must think it queer
> To stop without a farmhouse near
> Between the woods and frozen lake
> The darkest evening of the year.
>
> He gives his harness bells a shake
> To ask if there is some mistake.
> The only other sound's the sweep
> Of easy wind and downy flake.
>
> The woods are lovely, dark, and deep,
> But I have promises to keep,
> And miles to go before I sleep,
> And miles to go before I sleep.

The student sees that, on the surface, the poem is simple. In fact, the title seems to say it all: a person stops by woods on a snowy evening. An algorithmic system might not get much more out of it than that, but the student can. The repetition in the last two lines tells her something. *And miles to go before I sleep.* She might interpret this to mean that the speaker in the poem has responsibilities to attend to before he can rest. Perhaps the poem is contrasting the peacefulness of the woods with the chaos of civilization. The speaker stands at the foot of the woods, marveling at its beauty and depth. What if he were to stay? We feel his reluctance as he decides to return to the daily grind of society.

She is able to read the poem in another way. Perhaps sleep and the woods refer not to a rest from responsibility, but to death. *The woods are lovely, dark, and deep.* Frost may have used the woods as a symbol of eternal rest. In this reading, what lies outside the woods is not the responsibility of living in society but the strain of life itself. Life is full of struggle, of failure, of suffering. But the woods are lovely, dark, and deep. They are an escape from the hardships of existence. What must one do to escape? The speaker contemplates suicide as an answer to his problems. However, in the final stanza, he realizes that there is much more he would like to accomplish in life. He cannot stay in the woods. He cannot sleep yet.

The student is able to interpret the lines of poetry in different ways. How does she do it? The process may not be that different from the way Westfall decides not to fire his weapon. First, she perceives the words on the page. Algorithms in her brain make connections between the words she reads and the knowledge she has acquired in life, as well as her emotions and conscious experience. These connections are passed to the conscious agent for further analysis. At a basic level, this transition provides her with an understanding of what the words mean. This is already more than the algorithmic system can do. It only processes the patterns of visual stimuli caused by the ink on paper. Upon reflection, however, the student can learn far more than the surface meanings of the words. She can find symbolism, motifs, and implications. She discovers these hidden meanings by contemplating the simple connections that have been sent from her unconscious mind. Algorithms tell her that, based on data stored in her memory, the word *sleep* is somehow linked to death. It is, however, up to her conscious agent to elaborate on that connection. She does so by drawing on the meanings of the other lines in the poem. The connection between sleep and death seems to be strong here when we consider the nature of the woods at night. They are dark, mysterious, perhaps even mystical, like death. The repetition of the lines gives her that foreboding sense that the speaker has contemplated death.

All sorts of connections are sent to the conscious mind. Most of them are insignificant, because algorithms cannot distinguish between what is meaningful and what isn't. There are many arbitrary connections she could consider when reading the poem. For example, take the lines:

My little horse must think it queer
To stop without a farmhouse near

The algorithm of her unconscious might connect the word *queer* to her understanding of homosexuality. The student, however, does not dwell on the possibility that stopping somewhere without proximity to a farmhouse is a homosexual act, or that any horse would have that belief. An equally insignificant connection might be detected when she reads these lines:

The only other sound's the sweep
Of easy wind and downy flake.

The student might recognize that Downy is the brand name of a certain fabric softener, yet she does not suddenly jump up and exclaim that Robert Frost is really talking about how his bed sheets are soft, like a blanket of snow. In her memory, the word *queer* has been related to homosexuality and the word *downy* to fabric softener, and these links are detected by the data-processing algorithms of the brain. However, that these connections are not relevant to understanding the depth of Frost's poem is for the student, as a conscious agent, to decide.

Westfall's construal of his dilemma and his selection of relevant concepts, as well as the interpretations the student comes up with for the poem, begin as simple connections made by algorithms. In my view, the same is true for our ideas in general. They begin with algorithm and are expounded upon by the agent. At this point, someone like Daniel Wegner might be inclined to suggest that a valid conclusion from my view is that, because our ideas are initiated by deterministic algorithms, we are not responsible for them. We should receive neither credit nor blame for anything we think up. Take Einstein's theory of relativity, for example. Must we say that Einstein deserves no credit for this scientific discovery because the idea began in his unconscious?

I do not think we need to say this, for two reasons. First of all, the idea only *began* in his unconscious, as a simple connection between fragments of data. The discovery came to be only after Einstein contemplated the connection, expounded on it, and developed it in his mind. The theory of relativity did not simply pop into his head. Second, remember that the connections made between bits of information in the algorithmic systems in the brain often evoke memories, emotions, and conscious experiences. Therefore, part of the reason why this very important data connection was made in Einstein's brain is that he thought a lot (to say the least) about the nature of physics. Eventually, a connection was detected between some of those thoughts and other data in Einstein's brain. Einstein then painstakingly reflected on the connection, playing it out in his mind, and so a theory was born.

In this sense, Einstein's consciousness affected the operation of his algorithmic unconscious. This is a phenomenon familiar not only to neuroscientists but also to us in our everyday experience. Neurobiologically speaking, we affect the algorithms in our brains all the time by creating and shifting synaptic connections. The plasticity of our neuronal network allows our thoughts and behavior to affect the way that this network processes incoming data. But we already know that. We know, for instance, that when we practice something, we become better at it. This is an example of how we, in a sense, program our brains. When a piano student begins a new piece, she has to exert herself consciously to play each note properly. Finding the proper fingering, playing each chord with the right touch, and creating flow between each musical phrase require intense concentration. With practice, however, the need for this mental focus fades, until eventually the student can play the song without giving it any thought. Her fingers seem to guide themselves.

What has happened here is that the conscious effort the student devoted to her study of the piece has, through synaptic plasticity, become integrated into the algorithm of her brain. Through thinking about and practicing the playing of piano, she has become a better pianist just as Einstein became a better physicist, and got the idea for the theory of relativity by thinking about physics.

It is well known that an adolescent raised in an environment of violence and desperation is likely to one day choose a life of crime. This, too, is reflected in our model of the algorithm-agent interaction. Just as in Einstein's brain, the information that the adolescent receives undergoes exhaustive algorithmic processing, during which connections are made between bits of data and certain conscious experiences. However, unlike in Einstein's brain, these connections are not about physics or the nature of the universe. They have to do with whatever the teen has perceived and thought about: theft, cocaine, gang warfare, brutality. As the moral agent deliberates about the future, links between experiences of this kind are common, if not the only ones that come to mind. Einstein developed the connections that came to mind into full-fledged ideas and theories about light and gravity and time. This young man develops the connections that come to his mind into ideas about crimes he can commit. The agent might be, in a sense, forced to choose a path of wrongdoing if that is all he knows.

This is not to say that the adolescent has no free will. When he acts criminally, he does so freely and consciously. Yet we might still be able to say that his responsibility for the act is diminished, if he has any at all. It depends on the case. Regardless, the model for the interaction between algorithm and agent is consistent with the idea that someone can act freely, without necessarily being morally responsible for the act. We consciously control our thoughts and actions, but we depend on the algorithmic machinery of the brain to process the data of perception and experience on which we base our decisions.

Conscious agency provides us with the ability and opportunity to construct a deep, internal understanding of the world, and from that understanding synthesize explanations, theories, and possibilities for our future. Finally, through reflection on these mental creations, the moral agent bestows on us the faculties with which to freely will our actions in a way consistent with our beliefs and moral vision. It is the algorithm of Westfall's brain that allows his detection of an explosion and the nearing of blurry figures in the distance, but it is his conscious mind that understands the significance of these perceptions and deliberates about the proper course of action. The interaction of these two systems, the first a computational, rule-based mechanism and the second a free, agent-controlled consciousness, allows us to transcend the bounds of determinism and achieve a wondrous kind of decision making. Yet another mystery remains. The biological machinery of the brain seems much like the former and not at all like the latter. We see no consciousness or moral agent in the brain, only the mechanics of an intricate, organic computer, run by the binary circuitry of neuronal networks. How is it that such a mechanism gives rise to the boundless, indetermined consciousness we have described? What evidence can we find to show that the agent is neither an illusion nor a ghost in the machine?

17

RISE OF THE MORAL AGENT

In 1910, in a letter to the woman he would eventually marry, Alfred Wegener wrote: "Doesn't the east coast of South America fit exactly against the west coast of Africa, as if they had once been joined? This is an idea I'll have to pursue."[1] Wegener also noticed that the coastlines of South America and Africa, though separated by the vastness of the Atlantic, exhibited similar kinds of geological structures. His research led him to find descriptions of the fossils discovered in both continents. Intriguingly, they were left by the same plants and animals, pointing to the possibility that, millions of years ago, the same organisms inhabited those areas at the same time. The prevailing theory at the time was that these regions were once connected by long, narrow land bridges, allowing organisms to travel in between the continents. The theory said that these bridges were later submerged beneath the rising water levels. Wegener, however, did not accept this explanation, and sought to develop a theory of his own.

What he proposed was the theory of continental drift, which evolved into our modern understanding of plate tectonics. The concept was revolutionary: the modern continents were at one time part of a single, enormous land mass. That, Wegener thought, was the reason that the continents seemed to fit together, and why identical fossils were found on the coasts of Africa and South America. Furthermore, Wegener suggested that the current geography of the world gradually developed as the continents drifted apart from one another. It was this part of his theory that caused him to be ridiculed by the scientific community.[2]

"If we are to believe [this] hypothesis, we must forget everything we have learned in the last seventy years and start all over again," said one scientist. Another simply called the theory "utter, damned rot!"[3] Geologists were unwilling to consider the possibility that the continents could move. There simply was no force, they assumed, that could cause the motion of a body that large. In their minds, the idea was absurd—certainly not something that a serious scientist should be wasting time on. Despite the tremendous amount of evidence Wegener provided, his theory was almost unanimously rejected among geologists, since no one was willing to consider the possibility that the land masses could move. They were content to try to solve the puzzle using only the ideas with which they were familiar. It was not until years later that new findings were released which supported Wegener's ideas—including his insistence that continental drift was plausible—and plate tectonics became the most widely accepted explanation of Earth's geography.

The physicist Niels Bohr faced similar opposition to his support of the new theory of quantum mechanics. Though this opposition was not as widespread as that which Wegener confronted, it was no less intimidating, especially as it was led by the most respected intellectual in the world, Albert Einstein, who believed the concept of quantum mechanics to be fundamentally flawed—actually, he thought it was just plain ridiculous. Bohr, with his colleague Werner Heisenberg, was arguing that electrons do not have fixed velocities or positions in space. We cannot ever know exactly where they are, only where they are *likely to be*, using calculations of probability. Quantum mechanics explains that the reason for this is that interactions between particles at the quantum level (the smallest level in physics) are not deterministic. They are *random*. Einstein found this premise utterly unacceptable. How could there be a physical fact about physical objects that is in principle unknowable? How could any physical object act outside of determinism? How could anything happen in the universe without a cause?

Einstein was not willing to think about the world as Bohr did. Every phenomenon known to scientists of any kind, up to that point, had been determined, part of the natural, causal framework of the universe. Why should we suppose that this quantum level is any different? How could the determinism of the world around us arise from indeterminism of tiny particles? Despite Einstein's outright rejection of indeterminacy, the theory of Bohr and Heisenberg triumphed in the scientific world, proven again and again through the years. Quantum mechanics continued to grow as a field, and, in this respect, Einstein was left behind.

In our time, the study of consciousness is potentially as revolutionary and as full of repercussions as the study of quantum physics was to Bohr's generation and the study of continental drift was to Wegener's. These three scientific pursuits also have something else in common: all of them have challenged prevailing scientific views.[4] Continental drift defied the scientific consensus that no force was strong enough to move the continents. Quantum mechanics attacked the idea that all physical phenomena are determined. Now, conscious agency threatens the idea that determined interactions could never give rise to something not determined.

With the exception of quantum mechanics, every natural process known to physicists has been determined. The research of Bohr, Heisenberg, and others showed that some processes in the world are indeterminate, meaning random. Events can happen without a cause. However, neither determinism nor randomness seems to be able to account for free will and moral responsibility. Recall the way we put this problem earlier:

1. If neurobiological determinism is true, everything we do is completely caused by prior biological events, so we cannot be held morally responsible for our actions.
2. If indeterminism is true, our actions are random and we cannot be held morally responsible for them.
3. Either neurobiological determinism or indeterminism is true.
4. Therefore, we cannot be morally responsible for our actions.[5]

In the above formulation I would argue that premise 3 is incorrect. It's true, of course, that, scientifically speaking, we have only studied events that are either determined or indetermined, but we have all also had the experience of using our minds to make expansive, boundless decisions and consciously willing our actions. We can agree that these processes do not seem to fall in either category. There must be a *third* category—not determined and not random—in which they belong. Since we are not dualists, this category cannot be one of nonphysical things. Consciousness must come from the brain somehow, and do so in such a way that it assumes the property of neither being determined nor random. The question is, how is that possible, when we know the brain itself is a determined, biological system?

I think that the answer lies in a special case of what's called *emergence*, or emergent properties, the idea that a system can be more than the sum of its parts. We gave several examples of this phenomenon earlier, such as the saltiness of sodium chloride and the greasiness of oil. These are emergent properties because they are completely different from the properties of each substance's constituent parts. Similarly, consciousness depends on the workings of the brain, though it has a very different set of features. So perhaps consciousness is an emergent property of the interactions of neurons in the brain just as saltiness is of the interaction of sodium and chloride.

This comparison falls short, however, when we consider the fact that, though the properties of sodium chloride are very different from those of individual sodium and chloride atoms, they are still determined by them. The compound that is created when sodium and chloride are bound will necessarily have a certain set of properties, one of which is saltiness. We, however, would like to say that consciousness emerges from neuronal interactions with a new set of properties *and* without being determined. It arises with a new kind of causal interaction. But is this kind of emergence possible? Can one causal system give rise to another? It can. It has happened before.

The physics of Isaac Newton, the physics we see occurring around us all the time, abides by deterministic laws. But where did this determinism come from? It arose from the nondeterministic, random interactions of quantum particles. *Determinism is an emergent property of randomness.*

Einstein and his followers believed that the entirety of physics consisted solely of determined processes. The work of Bohr and Heisenberg revealed that our understanding of physics had to be expanded to include another kind of process: quantum mechanics. Now there are also those who believe that physics should be, and eventually will be, expanded to include a third process, one that explains consciousness. Just as the determinism of Newton's physics arose from the randomness of Bohr's, the agency of the mind emerged from the determinism of the brain.

It is even possible that the mind emerged from the interaction of random and determined processes. Some scientists, for example, have pointed out that brain activity is riddled with chaotic operations. To make a long story short, chaotic functions tend to magnify small differences in incoming data—making it more likely that improbable events will occur. It is therefore likely that these functions magnify quantum effects, leading to a greater expression of indeterminism in the brain. It could be that free will and conscious agency arise when this indeterminism interacts with deterministic neural processes. This is what might allow our minds to be free of the constraints of algorithmic processing.

The unique, nonalgorithmic nature of consciousness may be the explanation for why scientists understand almost nothing about it. For one thing, it is harder to study the operation of a system when one cannot make accurate predictions about its behavior. Psychologists study human behavior, but they cannot accurately predict it all the time. They cannot write out a rulebook based on human personalities that models how we will act, since, of course, human decision making is not rule based. Free will and agency, which may arise from the deterministic and indeterministic interactions of the brain, leave open the possibility that we will act in unpredictable ways.

It seems to me that another reason that consciousness remains beyond scientists' understanding has to do with their approach. Scientists, in general, treat conscious events as if they were any other determined bodily process. Thoughts, decisions, and willful acts are biological operations no different from the pumping of oxygenated blood from the heart or the release of digestive enzymes in the stomach. If mentality is regarded in this way from the outset, how can we expect to make progress in unraveling its mysteries?

To confront the extraordinary challenge posed by the study of consciousness, we may need a new approach, one designed to elucidate the nonalgorithmic, nonrandom aspects of its workings. This could mean an expanded understanding of physics. In Bohr's generation, the field was broadened to account for the interactions of quantum particles. Perhaps it should be expanded once more to include the interactions of thoughts. As Nobel Prize–winning physicist Eugene Burns has said: "Present-day physics represents a limiting case—valid for inanimate objects. It will have to be replaced by new laws, based on new concepts, if organisms with consciousness are to be described."[6] What is certain is that we should not repeat the mistake that was made in Wegener's day. The scientists who reviewed his concept of continental drift were quick to dismiss it because they could not imagine the possibility of forces strong enough to move the continents. Their assumptions of what constitutes utter implausibility held them back from contributing to the development of a momentous step forward in geology.

Of course, this new approach may not require a scientific revolution on the scale of Wegener's. A biological account of consciousness need not force us to revise the foundational principles of physics. Perhaps scientists already have all the theoretical tools they need. If this is the case, then the key to progress is for scientists to recognize that conscious agency is a real phenomenon that can be studied scientifically.

Consciousness is unlike anything else that has been subjected to scientific scrutiny. As such, we might expect it to require unique theories and innovative kinds of analysis, and the explanation that fits best must not be cast aside simply because it is unfamiliar, as was the case with Wegener's theory of continental drift. In the case of free will and agency, the unfamiliar may well be the most plausible account. We know that our actions cannot be determined, because determined systems are unable to address the boundless problems that we grapple with every day of our lives. Yet our decisions also cannot be random and arbitrary. A random system would not be able to match the depth of human reasoning. It would not be able to reason at all. What seems most plausible is to say that we, as moral agents, have conscious control of our actions, and that this agency somehow emerges, in a way that is not determined, from the interactions of neurons in the brain.

Whether nondeterministic emergence is what explains the rise of moral agency, it is a theory that respects the special attributes of consciousness that we see demonstrated again and again in human behavior. In the coming years, I have no doubt that today's theories of conscious agency will be refined and new ones proposed. A nonalgorithmic explanation of human thought will be achieved.

For this to occur, we will have to begin asking the right questions— questions driven by a commitment to explain free will rather than explain it away. It is only when we open ourselves to such a view that the scientific search for agency can begin. The goal should be to conceive of an approach that appeals to the special and profound nature of the mind. Only then can we begin to uncover its secrets.

18

THE PALACE OF THE MIND

Centuries ago, a nobleman was hosting a party at his palace for a crowd of guests. After seating everyone and directing his aides where to go, he decided to step outside for a breath of fresh air. It was then that a powerful gust of wind blasted through the palace, smashing the walls and supporting columns. Within moments, the roof of the palace began to crumble, and it then collapsed upon those seated in the hall.

Bodies lay everywhere, mangled and contorted. Peering at the wreckage, families of the deceased could not recognize their loved ones and could not identify them for burial. But the nobleman had a way. In his mind, he navigated the halls of his palace that once stood. He pictured the way he assigned the tables, the areas to which he sent his aides. He recalled his brief conversations with each guest, the walls on which they were leaning, the gentleman who asked to be pointed toward the restroom, the couple seated in the corner. By imagining himself walking through the palace and recalling the orientation in which his guests were seated, the nobleman was able to discover the identities of the bodies lying among the rubble and give their families some peace of mind.[1]

In the sixteenth century, the scholar Matteo Ricci told this story to a number of Chinese representatives in order to introduce them to a new strategy for recalling information: the *memory palace*. The idea is that a person can organize an incredible amount of information in his mind by imagining a grand palace in which each room contains a number of concepts. About 1,500 years earlier, a scholar of rhetoric named Quintillian described how one might construct a memory palace:

> The first thought is placed, as it were, in the forecourt; the second, let us say, in the living room: the remainder are placed in due order all [around] ... and entrusted not only to bedrooms and parlours, but even to the care of statues and the like. This done, as soon as the memory of the facts requires to be revived, all these places are visited in turn and the various deposits are demanded from their custodians, as the sight of each recalls the respective details. Consequently, however large the number of these which it is required to remember, all are linked one to the other like dancers hand in hand, and there can be no mistake since they join what precedes to what follows, no trouble being required except the preliminary labour of committing the various points to memory. What I have spoken of as being done in a house, can equally well be done in connection with public buildings, a long journey, the ramparts of a city, or even pictures. Or we may even imagine such places to ourselves.[2]

With this mental construct, Ricci said, the agent can, in the midst of his deliberation, navigate the palace of his thoughts and easily access any of the memories and experiences he finds relevant to his situation. He can go from room to room as he pleases, and even create or destroy rooms. The palace can be remodeled at will, or even transformed into a townhouse or office building. The agent can move through, reorganize, or alter his inner world of thoughts as he pleases. As historian Jonathan Spence describes it:

> He stands on the threshold of the Memory Palace, in his embroidered shoes.... In front of him, as far as the mind can travel, stretch the gleaming walls and colonnades, the porticoes, and the great carved doors, behind which are stored the images born of his reading, his experience, and his faith.... He feels the touch of cheek on cheek as the dying Francesco de Paris throws his arms around his neck.... Behind him wait two women, each cradling a child in her arms.... Through the quiet air, confusedly, comes the murmur of sound from the streets of Peking. He closes the door.[3]

Matteo Ricci used the memory palace as a strategy for organizing concepts, but it also serves as a metaphor for the working of conscious agency. The moral agent navigates the inner world of his thoughts. He can travel down any hall he chooses. He goes from one room, one set of concepts, to another, reflecting on the meaning and importance of each. He can rearrange the contents of any room, even move items from one room to another. The conscious mind has structure and organization.

However, at the same time, the palace of the mind is not bounded by definite walls. In your exploration of the mind's palace, you might discover secret passages—connections between rooms that you never knew existed. When confronted by, say, a difficult moral dilemma, rooms can be added, expanded, reduced, or destroyed. The limits of the palace can always be pushed. Human consciousness is boundless. It is this ability to navigate the depths of the mind that allows us to have introspective moral reasoning.

When Stephen Mobley finds himself pointing a gun at the manager of a pizza store, he frames the situation in the context of his experience. He construes the set of stimuli presented to him—the gun, the bag of money in his hand, a man on the floor crying—as a volitional problem with moral and practical implications.

Navigating the halls of his mind, Mobley considers the practical implications of letting his victim live versus killing him. The store manager could call the cops. He could describe Mobley's appearance to them and assist in his identification and capture. On the other hand, eliminating him would make him guilty of armed robbery *and* murder, if he were to be caught. He might even receive the death penalty.

As for the moral implications, Mobley may or may not have consciously weighed them at the time, but he was certainly *aware* of them. Neurologically healthy as he was, he had no problems distinguishing right from wrong. Standing in his mental palace, the hallway of moral reasoning was open to him. Either he walked down it and decided to ignore what he discovered or he refused to enter the hall altogether. He exercised his conscious will accordingly.

Mobley's lawyers would have argued that he couldn't have engaged in any sort of moral deliberation because he acted out of impulse. He pulled the trigger in the heat of the moment. However, even if we were to grant that this is true, that does not exonerate him. When he consciously decided to walk into a pizza store with a firearm, he knew exactly what he was getting into.[4] Having had experience with criminal behavior, Mobley was intimately familiar with his tendencies under pressure. The robbery was premeditated. Before he even entered the store, he considered what might happen inside. It wouldn't have taken much thought to realize the likelihood that he was going to leave a murder victim behind.

Regardless, there was nothing in Mobley's brain, before he entered the store or even while he had his gun pointed at the store manager, that prevented him from utilizing his power of moral agency and making the choice he thought he should make. Mobley's brain did not make him murder. He chose to do that himself.

Mobley is a moral agent, like the many others we have considered: Jean Valjean, Jean Bauby, the CIA station officer, the investigator of the bus accident, and Charles Westfall. What connects these individuals is that which occurs in their minds as they deliberate. Not enough is known about this process to be able to explain it with any level of scientific precision, but we can point out its essential attributes. We already know what they are.

The moral problems of Mobley, Valjean, Bauby, the CIA officer, the investigator, and Westfall are each contemplated by an *agent*: a unified, conscious entity with the power of free will. The agent in each case reflects on a wealth of background knowledge, an *internal world of experience*. The ability of the agent to navigate this world, as one might navigate a grand palace, is what makes human reasoning nonalgorithmic. The ability of the agent to expand this world—to alter it, transform it, and push its limits— as one might reorganize, or add rooms to, a palace is what makes human reasoning boundless.

It is what allows Valjean to consider the many facets of his dilemma, ranging from his promise to the bishop that he will be an honest man to the obligation he feels toward the citizens of Montreuil-sur-mer as the head of a major industry. It is what allows Bauby to explore the depths of his mind, conceptualize his struggle with locked-in syndrome, find the will not to let his mind wither away, and express himself in words, using a form of communication with which he had no prior exposure. It is what allows the CIA station officer to make both moral and practical considerations when ruminating over the ethics of secret missions. It what allows the bus investigator to explain an event using disconnected fragments of evidence, as he concludes from the discovery of an empty vodka bottle that the driver is lying. Finally, it is what allows Charles Westfall, suspecting danger in his midst, to contemplate his situation, choosing the most relevant aspects of what he sees and hears, in order to create a comprehensive mental picture of the events around him and decide on what he believes to be the moral course.

Each of these moral agents uses the powers endowed by consciousness to navigate the palace of his mind, his inner world of experience, to discover what he believes to be the best resolution to a problem, and will himself to implement it.

As we have seen, a determined system can try to imitate moral behavior, but it cannot truly achieve that which we as moral agents can. The processing of a determined system is governed by a set of rules from which it cannot deviate. In no sense can it freely navigate the store of information it collects, nor can it choose to expand or change the way it functions without that change being determined by its set of rules. It cannot solve boundless problems, such as that of Valjean or Westfall, because it is limited by the rules of its algorithmic processing. This is, of course, not to say that such a system would be unable to generate a result to a moral question. Recall that the ethics program used by the CIA station officer was able to mathematically weigh the variables of the scenario and return an answer that the Accordion mission should be aborted.

However, this result alone is not moral deliberation. On the inside, the program is just running a set of calculations, completely unaware of what they represent. Whether or not the program's answer is ethically correct is irrelevant, because it is not the result that we are concerned with. The essence of moral introspection is the *process*. Had the station officer made his choice based on a coin flip, he might not have authorized the mission either, but it would have been clear that he did not engage in any sort of moral reasoning to reach that conclusion. The station officer, like Valjean, Bauby, and Westfall, is a moral agent because his decision, whatever its resolution, is based on freely controlled, conscious introspection. His computer program, like any determined system in nature, just follows a set of rules, without ever making a single moral choice.

When a person with Tourette's syndrome insults someone because of an unconscious tick, it's clear that the act was caused by irregular interactions in his brain, not by an emergent, conscious agent. This is why he, like Heracles under Hera's curse, is not morally responsible for the act. Since conscious introspection cannot effectively wield control over his brain, his behavior is generated by a determined system.

Throughout this book, we have looked at examples of moral decisions and shown how human beings can work through them while determined systems cannot. But why use so many moral examples? Boundless problems certainly don't have to be ethical. The simple answer is that we have been investigating how our having *moral* responsibility depends on our having free will, so it just seems appropriate to use moral examples. However, there is a deeper reason.

In the branch of philosophy known as *metaethics*, the primary objective is to define what *good* is. Philosophers want to know what people mean by the terms *moral* and *immoral*. If we say that an act is morally good, what do we really mean? In all the literature that has been written on the subject, the clearest theme is that, strangely enough, we can't precisely define the terms *good* or *moral*. They are concepts that we simply have to get a sense of through experience.

Moral decisions are all about pursuing that which is morally good. However, *we can't even precisely define what is good*. So at the core of every moral decision is an ambiguity, one that cannot be resolved through any mathematical equation. Equations require that all terms be strictly defined. Countless books have been written that try to establish what we mean by *good, evil, moral,* and *immoral,* and there still is no consensus. Moral problems are, at their essence, boundless problems, adapted uniquely for the boundless reasoning of the human mind.

That our minds are well suited for moral decision making may be more than just coincidence. When consciousness emerged in the brain, many things changed. Personal identity and agency developed, where before there was only raw mechanism. As conscious beings, we are aware of the world and of ourselves. We are cognizant of our emotions, mindful of our beliefs and desires. Most important, however, we are aware of our freedom and the wealth of choices within our grasp. Confronted with so many possible courses of action and the power to navigate our minds to choose among them, we have the ability to ask the question that a determined system cannot ask: "What ought I to do?" I think that the ability to consider how we should act led us to ask the first moral questions.

Thought of in this way, our having conscious agency is a foundation of morality. Determined, mindless systems cannot deliberate or judge the nature of the behaviors they generate. The ability to reflect on the ethics of our actions is the gift, or perhaps the burden, that comes with having a mind.

We can now see most vividly why moral responsibility rests on our capacity for free will. Without the conscious ability to control our actions, *there would be no morality.* Ethical questions arose because we gained the ability to question ourselves, to reflect on our mistakes and successes, and choose a better future based on the wisdom gained from our experience. The mind is what endows us with moral responsibility and, save for those whose minds are unhealthy, we all must use the power bequeathed to us to make the best decisions we can, for ourselves and our fellow conscious agents.

Our ascendancy over the algorithm and mechanism of our biological construction is a triumph of evolution.[5] In addition to morality, things like creativity, introspection, friendship, art, philosophy, and society would be lost if it were not for the existence of consciousness. However, this is not to say that consciousness is better than algorithm in every respect.

For one thing, when confronted with an algorithmic problem, one definable as a strict set of rules, human beings will always be defeated by machines. The raw calculating power of a computer, which can be increased exponentially with better technologies, is, in many cases, vastly superior to the mathematical capacities of any person. A cheap, plastic, four-function calculator can do arithmetic better than any mathematician in the world.

The mind has another drawback. Every healthy person has the capacity for moral agency and an inner world of experience in which the agent navigates. The interaction between the agent and his conscious experience can be used in different ways—not all of them positive. For example, our capacity for self-questioning, which arises from that interaction, poses the risk of *excessive* reflection. The dangers of this possibility are illustrated most vividly in the eccentric narration of Dostoevsky's protagonist in his book *Notes from Underground.* The character, whom we call the Underground Man, believes that consciousness is a weakness of human beings, a terrible vulnerability that causes the thinking man to be paralyzed by doubt and indecision, and even to engage in self-harming behavior. At each moment, there are an immeasurable number of choices available to us. For each one, we can imagine a seemingly endless tree of consequences. We can formulate arguments in favor of each course, but we inevitably question our arguments and question our questioning. In our self-reflection, we find reasons to doubt our every thought and intention. The reasonable initiation of action soon becomes impossible. This is why the Underground Man suffers from his conscious deliberation. He says, "Consciousness, in my opinion, is the greatest misfortune for man, yet I know that man loves it and would not give it up for any satisfaction."[6] He believes that the reflective man tortures himself with overthinking. The only people free of this burden, says the Underground Man, are the stupid ones, those who act without thinking:

> Such a man simply rushes toward his object like an infuriated bull with its horns down, and nothing but a wall will stop him.... I envy such a man until I am green in the face. He is stupid. I am not disputing that, but perhaps the normal man should be stupid, how do you know? Perhaps it is very beautiful in fact.... Take the antithesis of the normal man, that is, the hyperconscious man.... [This] man is sometimes so nonplussed in the presence of his antithesis that he genuinely thinks of himself as a mouse and not a man.[7]

Of course, we would probably not agree that it is better to be foolish than thoughtful, but the message of the Underground Man can still resound with us. With the extraordinary advantages of self-reflection—the boundless exploration of the mind's palace—come the risks of self-doubt and self-harm. But that is all part of what it is to be human. We strive to use our freedom of choice to the best of our ability, but in the end we are all imperfect creatures. A moral agent might still do what he knows to be immoral in order to pursue more selfish ends. With his free will he can act unpredictably. He can decide to shock us by being spiteful or vengeful, or perhaps surprise us with an unexpected act of caring. The agent can navigate his inner world of experience in any way he chooses, and, for better or worse, this is what distinguishes us from machines and from animals, and what makes us who we are.

As far as we may have come in our investigation of consciousness, free will, and moral agency, we still don't know how the brain makes it all possible. This is perhaps the greatest scientific mystery of all time. The study of human thought is the new frontier of neuroscience, with the cooperation of dozens of other disciplines, and is still in its infancy. However, as more and more scientists begin to wonder about the great unknowns of brain function, the focus on the mind is intensifying.

Our understanding of moral agency and its attributes, however basic, provides the questions with which to start the investigation. How do scattered neuronal interactions give rise to unified human agency? How does the moral agent draw upon a seemingly endless internal world of experiences, emotions, and ideas in order to make decisions, even those about abstract concepts like morality? These are the questions facing neuroscientists today. They are broad, for now, but once we begin grappling with them using the technology and methodologies of modern neuroscience, they will become more refined, more targeted. In my view, it is by directing brain research on this path that we will eventually understand how the mind works, how the palace of the mind is constructed from the bricks and mortar of neurotransmission. The questions are fascinating, and the intensive pursuit of the answers is just beginning.

I said earlier that the emergence of consciousness was a triumph of evolution. Since that point, conscious human beings have used their powers of reason and contemplation to discover many things about the world—the genetic code, the world of bacteria and viruses, the nature of space and time—but I think the greatest discovery of all has yet to be made. I believe that the greatest triumph of consciousness will be to achieve a complete understanding of itself, to learn how the palace of the mind is created and how the conscious agent navigates and maintains it. Discovering the nature of our own thinking is the ultimate boundless problem, the ultimate challenge for experience. In that sense, the problem of consciousness was made for us to solve. And I think we are up to the task.

FURTHER READINGS

T his book deals with a lot of ideas that can't all be classified neatly under a single heading. For this reason, it could be difficult to know where to look if one wants to delve more deeply into the topic. As I was doing research for this book, I always found that, when I went to the library, no two references I needed were located in the same aisle. They were spread between the literature of philosophy, biology, neuroscience, computer science, psychology, politics, and others. I found this intriguing because it reflected the vastness of the problem and its relevance to many disciplines, but I would have to agree that it isn't convenient. So in what follows I will suggest what I think are good books to help take further an exploration that we have merely begun.

CHAPTER 1: THE NEFARIOUS NEURON

As good primers on the philosophy of mind, I would recommend two books by philosopher John Searle: *Mind: A Brief Introduction* and *The Mystery of Consciousness*. The latter might prove especially interesting because it includes exchanges between Searle and other philosophers. A good collection of influential essays is *Philosophy of Mind: Classical and Contemporary Readings*, edited by David Chalmers. For a broad introduction to the debate on consciousness, including a discussion of robots and artificial intelligence, see my earlier book, *Are You a Machine? The Brain, the Mind, and What It Means to Be Human*. Finally, Fyodor Dostoevsky's novel *Notes from Underground* is filled with references to, and remarkable insights into, free will and the nature of consciousness.

CHAPTER 2: THE SHADOW OF DETERMINISM

For a good discussion of the implications of determinism, try *How Free Are You? The Determinism Problem* by Ted Honderich. There are hundreds of books on quantum mechanics. To learn more about the role of quantum mechanics in nature, one good one is *The Physicists' View of Nature: The Quantum Revolution* by Amit Goswami. Another excellent book to check out on quantum mechanics is *The Quantum Brain: The Search for Freedom and the Next Generation of Man* by Jeffrey Satinover. Santinover not only gives a good introduction to the scientific principles behind the quantum but also connects it to the debate over the existence of free will.

CHAPTER 3: THE ESSENTIAL FREEDOM

The philosophical literature on free will is endless, but you can get a broad sense of the controversy by looking at collections of major papers. One that I would recommend is *The Oxford Handbook of Free Will* by Robert Kane. For something a little more concise, I would recommend the book *Free Will* by Daniel O' Connor. It isn't a collection, but it is a great introduction to the subject. If you are interested in exploring the relationship between free will and responsibility, try *Free Will and Moral Responsibility* by Peter French.

CHAPTER 4: A TEMPEST IN THE BRAIN

The idea of the moral agent, or the *self*, goes back a long way in philosophy. This book did not address any of that philosophy, at least not directly, but it is a fascinating literature worth looking into. To get started, a good collection of readings is *Self and Subjectivity* by Kim Atkins.

We, however, focused on a more scientific perspective on agency. Some books that I like on this topic are *The Astonishing Hypothesis: The Scientific Search for the Soul* by Francis Crick, *Synaptic Self: How Our Brains Become Who We Are* by Joseph LeDoux, *How the Self Controls Its Brain* by John C. Eccles, and *I Am a Strange Loop* by Douglas Hofstadter. To learn more about the relation between emergent properties and consciousness, look for the works of Gerald Edelman, such as *A Universe of Consciousness: How Matter Becomes Imagination* and *Wider Than the Sky: The Phenomenal Gift of Consciousness.*

CHAPTER 5: NEUROLOGICAL DISTURBANCE

Studies of patients with neurological injury or disease have provided scientists with many insights into the workings of the brain. One area of neuroscience that very much depends on this research is that of brain plasticity. An excellent book on this subject that explores dozens of clinical cases and their implications is *The Brain That Changes Itself* by Norman Doidge. To learn about the brain and consciousness at a more neuronal level, try *The Quest for Consciousness: A Neurobiological Approach* by Christof Koch and *Essential Sources in the Scientific Study of Consciousness*, edited by Bernard Baars and his colleagues. For a more general introduction to how the brain works, a very accessible book is *Mind Wide Open: Your Brain and the Neuroscience of Everyday Life* by Steven Johnson.

CHAPTER 6: THE SEAT OF THE WILL

To discover the true seat of the will would unlock a wealth of research opportunities. How does our initiation of behavior connect to the operation of the brain? As we discussed in this chapter, the search for the will seems to lead us to the frontal lobe. To learn more about the frontal lobe, and its relation to executive function, look for *The Executive Brain: The Frontal Lobes and the Civilized Mind* by Elkhonon Goldberg. Also relevant is Francis Crick's *The Astonishing Hypothesis: The Scientific Search for the Soul* and

Jeffrey Satinover's *The Quantum Brain: The Search for Freedom and the Next Generation of Man.*

CHAPTER 7: THE SOMATIC-MARKER HYPOTHESIS

To learn more about the somatic-marker hypothesis, read *Descartes' Error: Emotion, Reason and the Human Brain* by Antonio Damasio. Other good books by Damasio include *The Feeling of What Happens: Body and Emotion in the Making of Consciousness* and *Looking for Spinoza.*

CHAPTER 8: THE READINESS POTENTIAL

Benjamin Libet's research on the timing of conscious experience has been hotly debated in the scientific literature. If you are interested in an extended discussion of his results, look for volume 11 of the journal *Consciousness and Cognition* (2002). Articles both supporting and attacking Libet's interpretations and experimental design can be found there. To learn more about Libet's research, see his *Mind Time: The Temporal Factor in Consciousness.* You can also check out his edited volume *The Volitional Brain: Toward a Neuroscience of Free Will,* which includes contributions from scholars in the fields of neuroscience, psychology, psychiatry, physics, and philosophy.

CHAPTER 9: THE GRAND ILLUSION

The contention that free will is an illusion has been developed by a number of philosophers, psychologists, and scientists. Two leaders of this school of thought that come to mind are Daniel Dennett, author of *Consciousness Explained, Breaking the Spell,* and *Elbow Room,* and Daniel Wegner, author of *The Illusion of Conscious Will.* Another book to check out is *Living without Free Will* by Derk Pereboom.

CHAPTER 10: NEURONAL DESTINY

In the scientific literature, the prediction of behavior using neuroscientific means is usually not explicitly linked to the discussion of free will. To follow up on the studies we looked at in this chapter, I would begin by looking online for relevant articles by neuroscientists such as Apostolos Georgopoulos. Also helpful are collections of articles such as *Neural Correlates of Consciousness* edited by Thomas Metzinger.

To learn more about how innovations in neuroscience, such as brain fingerprinting, might affect law and the justice system, check out *Brain Policy: How the New Neuroscience Will Change Our Lives and Our Politics* by Robert Blank and *Neuroscience and the Law: Brain, Mind, and the Scales of Justice* edited by Brent Garland.

CHAPTER 11: THE REVOLUTION OF THE BRAIN

For more on the place of neurobiology in technological development, look for *The Neurotransmitter Revolution: Serotonin, Social Behavior, and the Law* by Roger Masters and Michael McGuire, *Mind Wars: Brain Research and National Defense* by Jonathan Moreno, and *From Morality to Mental Health: Virtue and Vice in a Therapeutic Culture* by Mike Martin. Also relevant is Norman Doidge's *The Brain That Changes Itself,* a book about the influence of environmental factors on brain functioning. To learn more about Prozac's effects on human personality, see Peter Kramer's *Listening to Prozac: The Landmark Book about Antidepressants and the Remaking of the Self.*

In writing this chapter, I got a lot of help from Richard Restak's book *The New Brain: How the Modern Age Is Rewiring Your Mind,* which I highly recommend. You might also enjoy two other of his books: *The Brain Has a Mind of Its Own: Insights from a Practicing Neurologist* and *The Naked Brain: How the Emerging Neurosociety Is Changing How We Live, Work, and Love.*

CHAPTER 12: SEEDS OF CORRUPTION

In recent years, the term *neuroethics* has arisen to describe the study of how neuroscience relates to morality. A good collection on this topic, from the perspectives of neuroscientists, philosophers, and ethicists, is *Neuroethics: Mapping the Field* edited by Steven Marcus. If you are interested in the relationship between neuroscience and the law, try *Brain Policy: How the New Neuroscience Will Change Our Lives and Our Politics* by Robert Blank, *Neuroscience and the Law: Brain, Mind, and the Scales of Justice* edited by Brent Garland, *The Neurotransmitter Revolution: Serotonin, Social Behavior, and the Law* by Roger Masters and Michael T. McGuire, and *The Naked Brain: How the Emerging Neurosociety Is Changing How We Live, Work, and Love* by Richard Restak. These all cover the issue of what neuroscience can teach us about criminal behavior. *The Neurobiology of Violence* by Jan Volavka is another good book on this topic, though it is more technical than the others. Finally, to learn more about the crime and trial of Leopold and Loeb, look for Hal Higdon's *Leopold and Loeb: The Crime of the Century.*

CHAPTER 13: MORALITY'S END

The majority of Gazzaniga's research can be found by searching the Internet for academic papers that he's written, but many of his opinions on issues like free will and morality are expressed in his book *The Ethical Brain.*

The work of Daniel Dennett is known to take a position similar to that taken in this chapter. Two good examples are *Consciousness Explained* and *Breaking the Spell.* You might also be interested in reading *The Myth of Morality* by Richard Joyce.

CHAPTER 14: THE DEPTHS OF CONSCIOUSNESS

Consciousness is truly a mysterious phenomenon. What is it? How does it come about? Since we cannot yet define it precisely, the best way to understand it is by considering its many attributes—the mental powers it endows. For a philosophical consideration of these various aspects, read

John Searle's *Mind: A Brief Introduction* and *The Mystery of Consciousness*. For a neuropsychological understanding, try *In the Theater of Consciousness: The Workspace of the Mind* by Bernard Baars. I also tend to enjoy collections of readings on consciousness because they give a variety of viewpoints and approaches to the subject that broaden my understanding. Two such collections are *Essential Sources in the Scientific Study of Consciousness* edited by Bernard Baars and his colleagues and *Neural Correlates of Consciousness* edited by Thomas Metzinger. For a more eclectic collection, which includes stories and essayistic reflections on consciousness, try *The Mind's I: Fantasies and Reflections on Self and Soul* edited by Douglas Hofstadter and Daniel Dennett. Finally, to learn more about the life of Jean Bauby, look for his inspiring book *The Diving Bell and the Butterfly*.

CHAPTER 15: A CHALLENGE FOR EXPERIENCE

In this chapter we explored arguments for the possibility that human consciousness cannot be algorithmic. We took the angle of showing how we can solve boundless problems, such as moral ones, but determined systems cannot. For a discussion of how this relates to the processing abilities of computers, look for *What Computers Still Can't Do: A Critique of Artificial Reason* by Hubert Dreyfus. For a detailed account of inductive reasoning and human problem solving, try *Better Reasoning: Techniques for Handling Argument, Evidence, and Abstraction* by Larry Wright.

The argument that algorithmic systems cannot solve boundless problems has also been addressed mathematically using a famous mathematical concept known as *Gödel's theorem*, which demonstrates that all such systems are limited in the kinds of problems they can solve. It is a fascinating proof and is applicable to many ideas of philosophy. For an introduction to its development and implications, look for *Gödel's Theorem: An Incomplete Guide to Its Use and Abuse* by Torkel Franzén. For a unique, however lengthy, discussion of the theorem's relation to everyday phenomena, such as art and music, you might also like *Gödel, Escher, Bach: An Eternal Golden Braid* by Douglas Hofstadter.

CHAPTER 16: THE BOUNDLESSNESS OF REASON

The bulk of the ideas in this chapter are my own, though they have been inspired by the work of philosophers John Searle and Hubert Dreyfus. To anyone who is interested in learning more about consciousness and human reasoning, I would especially recommend Searle's work because it is very clearly explained. Two of his books that come to mind are *The Mystery of Consciousness* and *Mind: A Brief Introduction*. You may also like *In the Theater of Consciousness: The Workspace of the Mind* by Bernard Baars. Finally, ideas relevant to those in this chapter are discussed in my earlier book, *Are You a Machine? The Brain, the Mind, and What It Means to Be Human*.

CHAPTER 17: RISE OF THE MORAL AGENT

In discussing how moral agency might have arisen, one body of literature that we did not cover is that on the evolution of consciousness. If you would like to tap into that body of research, a start would be *A Brief History of the Mind* by William Calvin.

For more on the status of science and how it relates conscious agency, try *The Quantum Brain: The Search for Freedom and the Next Generation of Man* by Jeffrey Satinover, *The Astonishing Hypothesis* by Francis Crick, and, if you'd like to get much more advanced, *The Emperor's New Mind* by Roger Penrose. To read about the emergence of moral agency, try the books of Gerald Edelman, such as *A Universe of Consciousness: How Matter Becomes Imagination* and *Wider Than the Sky: The Phenomenal Gift of Consciousness*. I would also recommend *I Am a Strange Loop* by Douglas Hofstadter.

CHAPTER 18: THE PALACE OF THE MIND

I believe that the neurobiological study of consciousness will bring about the next scientific revolution. Besides all the cutting-edge research going on, the questions of neuroscience themselves are fascinating and truly worth a bit of reflection. If any of the ideas in this book have captured you, here are a few others that you might like: *The Quest for Consciousness: A Neu-robiological Approach* by Christof Koch, *Essential Sources in the Scientific Study of Consciousness* edited by Bernard Baars and his colleagues, *Mind Wide Open: Your Brain and the Neuroscience of Everyday Life* by Steven Johnson, *The Brain That Changes Itself* by Norman Doidge, *Neural Correlates of Consciousness* by Thomas Metzinger, *The Astonishing Hypothesis: The Scientific Search for the Soul* by Francis Crick, *The Mind's I: Fantasies and Reflections on Self and Soul* edited by Douglas Hofstadter and Daniel Dennett, and my own *Are You a Machine? The Brain, the Mind, and What It Means to Be Human.*

More on the idea of the Memory Palace can be found in *The Memory Palace of Matteo Ricci* by Jonathan Spence. If you are interested in learning about the road to breakthroughs in neuroscience research, you should check out *In Search of Memory: The Emergence of a New Science of Mind* by Eric Kandel, winner of the Nobel Prize in Physiology or Medicine for his work on the biological basis of memory. There are also hundreds of other books out there on consciousness, free will, and moral agency, if you are interested in going further.

NOTES

INTRODUCTION

1. Fyodor Dostoevsky, *Notes from the Underground*, trans. Ralph E. Matlaw (New York: E. P. Dutton and Co., 1960), p. 24 [emphasis added].

CHAPTER 1: THE NEFARIOUS NEURON

1. *Mobley v. State*, 455 S.E. 2d 61 (Ga. 1995).

2. Cecille Price-Huish, "Born to Kill? Aggression Genes and Their Potential Impact on Sentencing and the Criminal Justice System," *Southern Methodist University Law Review* (January 1997): 610.

3. Francis Crick, *The Astonishing Hypothesis: The Scientific Search for the Soul* (New York: Charles Scribner's Sons, 1994), p. 3.

4. Joseph LeDoux, *Synaptic Self: How Our Brains Become Who We Are* (New York: Penguin Books, 2002), p. 324.

5. As quoted in Dennis Overbye, "Free Will: Now You Have It, Now You Don't," *New York Times*, January 2, 2007.

6. The argument here has been represented in a philosophical structure to make it easy to see the premises and conclusion. This structure makes it easier for us to consider which aspects of the argument we agree with and allows for targeted critiques of which premises we find problematic.

7. John R. Searle, *Mind: A Brief Introduction* (Oxford: Oxford University Press, 2004), p. 160.

CHAPTER 2: THE SHADOW OF DETERMINISM

1. "He listens to the Brain's 'Sur,'" May 29, 2001, http://www.rediff.com/news/may401us.htm.

2. "Expanding *Nature Neuroscience*," *Nature Neuroscience* 8 (2005): 1.

3. Enoch Gordis, as quoted in Sandra Ackerman, *Discovering the Brain* (Washington, DC: National Academy Press, 1992), p. 7.

4. Pierre Laplace, *A Philosophical Essay on Probabilities*, trans. F. W. Truscott and F. L. Emory (New York: Dover, 1951).

5. This short story, called "Appointment in Samara," was written by W. Somerset Maugham in 1933.

6. A *quantum* is defined as the smallest possible unit of something. A quantum of light, for example, is a photon. Quantum mechanics is the study of physics at the smallest possible scale.

7. Amit Goswami, *The Physicists' View of Nature: The Quantum Revolution* (New York: Springer, 1992), p. 55.

8. In constructing the relation this way, I rely on Russ Shafer-Landau and Joel Feinberg, *Reason and Responsibility: Readings in Some Basic Problems of Philosophy* (Belmont, CA: Wadsworth, 2004), pp. 387–88.

CHAPTER 3: THE ESSENTIAL FREEDOM

1. Known in Roman mythology as *Hercules*.

2. It seems clear that one cannot be held responsible for an act that was not free. However, could someone do an act freely for which he is not morally responsible? Suppose that, instead of causing him to kill his family, Hera tells Heracles to steal a chest of gold for her, threatening to kill him if he does not. Using his free will, Heracles does what Hera asks. Does that mean that he is morally responsible for taking the chest? Maybe not. To answer this question, we would need a lot more information about the ethical issues involved. The point, however, is that Heracles may not be morally responsible, even though he uses his free will. We see from here that there are situations in which a person acts freely but isn't morally responsible. In short, free will—if it exists—may not be sufficient for moral responsibility, but it surely seems necessary for it.

3. Compatibilism is sometimes referred to as *soft determinism,* a term coined by psychologist William James. Because I find the term misleading and not very useful, it will not be used in this book.

4. A thought experiment like this one was originally suggested by John Locke in *An Essay Concerning Human Understanding.* This kind of example is commonly referred to as a *Frankfurt-type example,* because it was used by philosopher Harry Frankfurt to challenge the idea that free will requires that one could have done otherwise.

5. Many have suggested this Frankfurt-style example, including Eleonor Stump, "Libertarian Freedom and the Principle of Alternative Possibilities," in *Faith, Freedom and Rationality,* ed. Daniel Howard-Snyder and Jeff Jordan (Lanham, MD: Rowman and Littlefield, 1996), pp. 73–88; and Stewart Goetz, "Frankfurt-Style Counterexamples and Begging the Question," *Midwest Studies in Philosophy* 29 (2005): 83–105.

6. John R. Searle, *Mind: A Brief Introduction* (Oxford: Oxford University Press, 2004), p. 154.

7. Ibid., p. 155.

8. Thomas H. Huxley, "On the Hypothesis That Animals Are Automata, and Its History," in *Collected Essays* by T. H. Huxley (Boston, MA: Adamant Media Corporation, 2000 [1874]), pp. 240–44.

CHAPTER 4: A TEMPEST IN THE BRAIN

1. Victor Hugo, *Les Misérables,* trans. Lee Fahnestock and Norman MacAfee (New York: Signet Classics, 1987), pp. 224–25.

2. Ibid., p. 227.

3. Ibid., pp. 230–31.

4. Ibid., pp. 219–34.

5. A *gyrus* is a protruding surface of the brain. This is to be contrasted from a *sulcus,* which is a line or crease located between gyri.

6. Exactly how all these brain structures work together to generate emotion is still unknown. For this reason, we cannot propose an accurate pathway of emotional processing.

7. This can be seen in the titles of many works in neuroscience, such as *Descartes' Error,* by neuroscientist Antonio Damasio (New York: Avon Books, 1994).

8. Gilbert Ryle, *The Concept of Mind* (London: Hutchinson & Company, 1949).

9. I recently heard an interesting story (I can't promise that it's true) about an attempt by a few people to prank their friend by dumping several pounds of sodium into his swimming pool. Fortunately for them, they were prevented from going through with it. It would not have been a very good prank because none of them would have lived to see the surprised look on their friend's face, since the pool, their friend's house, and much of the area around it would have been transformed into a giant fireball.

10. See Gerald Edelman, *A Universe of Consciousness: How Matter Becomes Imagination* (New York: Basic Books, 2000); and Christof Koch, *The Quest for Consciousness: A Neurobiological Approach* (Englewood, CO: Roberts and Company, 2004).

CHAPTER 5: NEUROLOGICAL DISTURBANCE

1. Elkhonon Goldberg, *The Executive Brain: The Frontal Lobes and the Civilized Mind* (Oxford: Oxford University Press, 2001), p. 182.

2. Sean A. Spence and Chris D. Frith, "Towards a Functional Anatomy of Volition," *Journal of Conscious Studies* 6 (1999): 11–29.

3. Ibid.

4. Gerald T. Lim et al. "Clinicopathologic Case Report: Akinetic Mutism with Findings of White Matter Hyperintensity," *Journal of Neuropsychiatry and Clinical Neurosciences* 14 (2002): 214–21.

5. Ibid., pp. 120–21.

6. Paolo Cavedini et al., "Frontal Lobe Dysfunction in Obsessive-Compulsive Disorder and Major Depression: A Clinical-Neuropsychological Study," *Psychiatry Research* 78, no. 1–2 (1998): 21–28.

7. François Lhermitte, "'Utilization Behavior' and Its Relation to Lesions of the Frontal Lobes," *Brain* 106, no. 2 (1983): 237–55; and Lhermitte, "Human Autonomy and the Frontal Lobes. Part I: Imitation and Utilization Behavior: A Neuropsychological Study of 75 Patients," *Annals of Neurology* 19, no. 4 (1986): 326–34.

8. Yutaka Tanaka et al., "Forced Hyperphasia and Environmental Dependency Syndrome," *Journal of Neurology, Neurosurgery and Psychiatry* 68, no. 2 (2000): 224–26.

9. Henrik Walter, "Neurophilosophy of Free Will," in *The Oxford Handbook of Free Will*, ed. Robert Kane (Oxford: Oxford University Press, 2002), pp. 565–76. The name "I-disorder" is a translation of the German term *Ich-Störung*.

10. A PET scan is taken by injecting a radioactive tracer, known as an *isotope,*

into a patient's bloodstream. As the isotopes decay, little particles called positrons are released from their atomic nuclei. What the scanner does is detect when the positrons collide with electrons. Since the tracer was put into the bloodstream, the scan effectively measures where in the brain the blood is flowing. It measures activation of the brain by reporting changes in regional cerebral blood flow (rCBF).

11. Sean A. Spence et al. "A PET Study of Voluntary Movement in Schizophrenic Patients Experiencing Passivity Phenomena," *Brain* 120, no. 11 (1997): 1997–2011.

12. The scientist who promotes determinism would probably disagree that Devere could not be held morally responsible because he is probably a compatibilist. He would say that Devere will always have free will and be morally responsible for his actions as long he had alternative choices. As established in the previous chapter, this is not the understanding of free will that we are discussing.

CHAPTER 6: THE SEAT OF THE WILL

1. Trevor H. Levere, *Transforming Matter: A History of Chemistry from Alchemy to the Buckyball* (Baltimore, MD: Johns Hopkins University Press, 2001), pp. 33–38, 56–69.

2. Oxidations are also defined as reactions that add an electronegative atom (such as oxygen) to or remove hydrogen from a molecule.

CHAPTER 7: THE SOMATIC-MARKER HYPOTHESIS

1. Here are a few measurements to give you an idea of what kind of rod this is. Weighing over thirteen pounds, it is three feet, seven inches long with a diameter of one and a quarter inches. See Antonio Damasio, *Descartes' Error* (New York: Avon Books, 1994), p. 6.

2. The account of Phineas Gage's accident and recovery was taken from Damasio, *Descartes' Error* (first chapter); and Michale S. Gazzaniga et al., *Cognitive Neuroscience: The Biology of the Mind* (New York: W. W. Norton, 2002), pp. 537–38.

3. This region is sometimes divided into two parts. When it is, the middle is referred to as the *ventromedial prefrontal cortex* and the right and left sides as the *lateral orbitofrontal cortex*.

4. Damasio, *Descartes' Error*, pp. 165–201.

5. Ibid, pp. 173–75.

6. There are bodily experiences associated with sadness, joy, anger, and the other emotions.

7. Damasio, *Descartes' Error*, p. 173.

8. See chapter 3 for several examples of how damage to the executive system affects behavior.

9. Damasio, *Descartes' Error*, pp. 35–44.

10. As part of this procedure, frontal lobe tissue had to be removed along with the tumor.

11. Damasio, *Descartes' Error*, p. 42.

12. It is not coincidental that this disorder sounds much like the word *perseverance*.

13. A way that an emotional dysfunction can be confirmed is by measuring skin conductance responses (basically, this is like a polygraph test) of patients with orbitofrontal damage to images that would generally elicit emotions. In Damasio's lab, orbitofrontal patients and normal controls were monitored as they viewed pictures of things like the Iowa countryside and a bloody corpse. See Gazzaniga et al., *Cognitive Neuroscience*, p. 552. Spikes were seen in the graphs of skin conductance for the normal subjects. The graphs for the patients, on the other hand, were near straight lines, indicating that they had almost no response to the emotional stimuli.

14. A. Bechara, A. R. Damasio, H. Damasio, and S. W. Anderson, "Insensitivity to Future Consequences Following Damage to Human Prefrontal Cortex," *Cognition* 50 (1994): 7–15; and A. Bechara, H. Damasio, and A. R. Damasio, "Emotion, Decision Making and the Orbitofrontal Cortex," *Cerebral Cortex* 10 (2000): 295–307.

15. Damasio, *Descartes' Error*, pp. 214–16.

16. Ibid, p. 173.

CHAPTER 8: THE READINESS POTENTIAL

1. In philosophy, this fallacy is sometimes called *ad hoc ergo propter hoc*, which is Latin for "after this, therefore because of this."

2. Since, as we established in chapter 3, moral responsibility cannot exist without free will.

3. Devoting a professional life to the study of consciousness is risky. Very little is known about it and quality results are difficult to come by. As Francis Crick (*Astonishing Hypothesis*) writes, Libet dared enter the study of consciousness only after he received tenure.

4. Benjamin Libet, *Mind Time: The Temporal Factor in Consciousness* (Cambridge, MA: Harvard University Press, 2004), pp. 33–34.

5. Kevan Martin, "Time Waits for No Man," *Nature* 429, no. 20 (2004): 243–44.

6. Libet, *Mind Time*, pp. 124–34.

7. Ibid., p. 134.

8. Ibid., pp. 137–40.

9. Ibid., p. 139.

10. Ibid., p 138.

11. Vilayanur S. Ramachandran, quoted in "The Zombie Within," *New Scientist*, September 5, 1998, p. 35.

12. Libet, *Mind Time*, p. 149.

13. Ibid., pp. 150–51.

14. Daniel C. Dennett, "The Self as a Responding—and Responsible—Artifact," *Annals of the New York Academy of Sciences* 1001 (2003): 39–50.

15. Benjamin Libet, "Do We Have Free Will?" in *Volitional Brain*, ed. Benjamin Libet et al. (Thoverton, UK: Imprint Academic,1999), pp. 47–58.

16. Judy A. Trevena and Jeff Miller, "Cortical Movement Preparation Before and After a Conscious Decision to Move," *Consciousness and Cognition* 11, no. 2 (2002): 162–90.

17. Libet replies by asserting that this was accounted for. He and his colleagues tried eliminating the earliest 10 percent of the measurements to see how the average would be affected. They determined that the average remained almost the same: about 350 milliseconds before the conscious decision to act. See Libet, "The Timing of Mental Events: Libet's Experimental Findings and Their Implications," *Consciousness and Cognition* 11 (2002): 291–99.

18. Timothy L. Hubbard and Jamshed J. Barucha, "Judged Displacement in Apparent Vertical and Horizontal Motion," *Perception and Psychophysics* 44, no. 3 (1988): 211–21; and Steve Joordens et al., "When Timing the Mind One Also Should Mind the Timing: Biases in the Measurement of Voluntary Actions," *Consciousness and Cognition* 11 (2002): 231–40.

19. Specifically, in parts of the medial frontal cortex.

20. Hakwan C. Lau et al., "On Measuring the Perceived Onset of Spontaneous Actions," *Journal of Neuroscience* 26, no. 27 (2006): 7265–71.

21. Trevena and Miller, "Cortical Movement Preparation Before and After a Conscious Decision to Move."

CHAPTER 9: THE GRAND ILLUSION

1. Daniel M. Wegner, *The Illusion of Conscious Will* (Cambridge, MA: MIT Press, 2002), pp. 341–42.

2. Ibid., pp. 64–98.

3. This example was adapted from ibid., p. 63.

4. Admittedly, I'm the one who threw in that line about the Macarena; Wegner's example only focuses on the movements of the branches. However, if we are going to play with this thought experiment (which is silly to begin with) it seems reasonable to say that a person would test the limits of her supposed power of "tree control" before concluding that she actually had it.

5. Daniel M. Wegner and Thalia Wheatley, "Apparent Mental Causation: Sources of the Experience of the Will," *American Psychologist* 54 (1999): 480–91.

6. René Descartes, *Meditations on First Philosophy*, trans. John Veitch in 1901. http://www.wright.edu/cola/descartes/.

7. Wegner, *Illusion of Conscious Will*, p. 49.

8. Especially since "yielding action" is such a vague idea. The term *system that yields action* is incredibly imprecise and I don't really know what it is supposed to mean.

9. Wilder Penfield, *The Mystery of Mind* (Princeton, NJ: Princeton University Press, 1975), pp. 76–77; and Wegner, *Illusion of Conscious Will*, p. 45.

10. Just to be clear, not all neurosurgeons are steeped in madness and obsessed with manipulating people's brains. Only some are.

11. Wegner, *Illusion of Conscious Will*, p. 11.

12. Frode Willoch et al., "Phantom Limb Pain in the Human Brain: Unraveling Neural Circuitries of Phantom Limb Sensation Using Positron Emission Tomography," *Annals of Neurology* 48 (2000): 842–49.

13. Ibid., p. 40.

14. Lynette A. Jones, "Motor Illusions: What Do They Reveal about Proprioception?" *Psychological Bulletin* 103 (1998): 72–86; and Wegner, *Illusion of Conscious Will*, p. 40.

15. See chapter 4 for a longer discussion of schizophrenia and other disorders of the will.

16. Wegner, *Illusion of Conscious Will*, p. 47.

17. Whether or not movements during hypnosis are actually determined is debatable.

18. John R. Searle, *Mind: A Brief Introduction* (Oxford: Oxford University Press, 2004), p. 157.

19. Wegner, *Illusion of Conscious Will*, p. 341.

20. Ibid., p. 342.

CHAPTER 10: NEURONAL DESTINY

1. "Solar Eclipses in History and Mythology: Historical Observations of Solar Eclipses," Bibliotheca Alexandria Online, March 29, 2006, http://www.bibalex.org/eclipse2006/HistoricalObservationsofSolarEclipses.htm.

2. Wil Milan, "Fear and Awe: Eclipses through the Ages," January 18, 2000, http://www.space.com/scienceastronomy/solarsystem/lunar_lore_000118.html.

3. "Solar Eclipses in History and Mythology."

4. Herodotus, *Clio*, http://www.greektexts.com/library/Herodotus/Clio/eng/329.html (accessed November 17, 2006).

5. He actually did the experiment using many different stimuli, such as the blowing of a whistle and the clicking of a metronome. In fact, he may not have used a bell at all, but that is how most people describe the experiment.

6. Apostolos P. Georgopoulos, "Neural Mechanisms of Motor Cognitive Processes: Functional MRI and Neurophysiological Studies," in *The New Cognitive Neurosciences*, ed. Michael S. Gazzaniga (Cambridge, MA: MIT Press, 2000), pp. 525–38.

7. This is not a uniquely neuroscientific concept. It is just the concept of vector summation in physics. For example, if two forces are applied to a box from different directions, the movement of the box will be the vector sum (magnitude and direction) of those two forces.

8. Of course, I do not mean "intention" literally in this case, only as part of the analogy.

9. Michael S. Gazzaniga et al., *Cognitive Neuroscience: The Biology of the Mind* (New York: W. W. Norton, 2002), p. 465.

10. Norman Doidge, *The Brain That Changes Itself: Stories of Personal Triumph from the Frontiers of Brain Science* (New York: Viking, 2007), p. 206.

11. Ibid., p. 207.

12. Ayelet Sapir et al., "Brain Signals for Spatial Attention Predict Performance in a Motion Discrimination Task," *Proceedings of the National Academy of Sciences* 103, no. 49 (2005): 17810–15.

13. Michael Purdy, "Researchers Use Brain Scans to Predict Behavior," Washington School of Medicine in St. Lewis Online, November 29, 2005, http://mednews.wustl.edu/news/page/normal/6248.html.

14. Brent Garland, *Neuroscience and the Law: Brain, Mind and the Scales of Justice* (New York: Dana Press, 2004), p. 105.

15. Ibid., pp. 103–106.

16. Ibid., p. 106.

17. My intention is certainly not to demean the results of this and related experiments. They are certainly very impressive. I want to show only that they do not demonstrate that *all* human behaviors are predictable.

CHAPTER 11: THE REVOLUTION OF THE BRAIN

1. Peter D. Kramer, *Listening to Prozac: The Landmark Book about Antidepressants and the Remaking of the Self* (New York: Penguin Books, 1997), pp. ix–xi.

2. Ibid., p. 15.

3. Richard Restak, *The New Brain: How the Modern Age Is Rewiring Your Mind* (New York: Rodale, 2003), p. 132.

4. Provigil is the brand name of a drug called modafinil.

5. "Narcolepsy More Common in Men, Often Originates in Their 20s," http://www.mayoclinic.org/news2002-rst/986.html.

6. Ibid.

7. Arlene Weintraub, "Eyes Wide Open," *Business Week*, April 24, 2006; and Restak, *The New Brain*, pp. 132–34.

8. Weintraub, "Eyes Wide Open."

9. Restak, *The New Brain*, p. 133.

10. Jonathan D. Moreno, *Mind Wars: Brain Research and National Defense* (Washington, DC: Dana Press, 2006), p. 116.

11. Sean E. McCabe et al., "Non-medical Use of Prescription Stimulants among US College Students: Prevalence and Correlates from a National Survey," *Addiction* 99 (2005): 96–106.

12. Restak, *The New Brain*, pp. 135–38.

13. Kramer, *Listening to Prozac*, pp. 1–12, quotation is from p. 12.

14. Johan A. den Boer, "Social Anxiety Disorder/Social Phobia: Epidemiology, Diagnosis, Neurobiology, and Treatment," *Comprehensive Psychiatry* 46, no. 6 (2000): 405–15.

15. Mark Barad et al., "Rolipram, a Type IV-Specific Phosphodiesterase Inhibitor, Facilitates the Establishment of Long-Lasting Long-Term Potentiation and Improves Memory," *Proceedings of the National Academy of Sciences* 95 (1998): 15020–25; Restak, *The New Brain*, pp. 141–44; and Anjan Chatterjee, "Cosmetic Neurology: The Controversy over Enhancing Movement, Mentation, and Mood," *Neurology* 63 (2004): 968–74.

16. From Judy Illes, as quoted in "Messing with Our Minds," *Independent*, January 18, 2005.

17. Chatterjee, "Cosmetic Neurology."

18. I address the question of whether human beings are machines in my previous book (published in 2007), titled *Are You a Machine?*

19. Restak, *The New Brain*, p. 147.

20. Ibid., pp. 138–40.

CHAPTER 12: SEEDS OF CORRUPTION

1. George P. Scott, *Atoms of the Living Flame: An Odyssey into Ethics and the Physical Chemistry of Free Will* (Lanham, MD: University Press of America, 1985), pp. 2–13.

2. Hal Higdon, *Leopold and Loeb: The Crime of the Century* (Champaign: University of Illinois Press, 1999), p. 17.

3. Ibid., p. 19.

4. Ibid.

5. It is debated who drove and who did the killing.

6. Higdon, *Leopold and Loeb*, p. 42.

7. Darrow is also known for his defense of the teaching of evolution in the famous 1925 Scopes Trial.

8. Higdon, *Leopold and Loeb*, p. 164.

9. Clarence Darrow, "A Plea for Mercy," in *Modern Eloquence*, vol. 6, ed. Ashley Thorndike (New York: P. F. Collier & Son, 1936), pp. 80–85.

10. Others have made the connection between this example and biological determinism. See, for example, Scott, *Atoms of the Living Flame*.

11. Clarence Darrow, *Crime and Criminals: An Address Delivered to the Prisoners in the Chicago County Jail* (Chicago: Charles H. Kerr and Company, 1919).

12. Dorothy Nelkin and M. Susan Lindee, *The DNA Mystique: The Gene as a Cultural Icon* (New York: W. H. Freeman, 1995), p. 144 [emphasis added].

13. Markku Linnoila et al., "Low Cerebrospinal Fluid 5-Hydroxyindoleacetic Acid Concentration Differentiates Impulsive from Nonimpulsive Violent Behavior," *Life Sciences* 33 (1983): 2609–14.

14. Ibid.

15. Matti Virkkunen et al., "CSF Biochemistries, Glucose Metabolism, and Diurnal Activity Rhythms in Alcoholic, Violent Offenders, Fire Setters, and Healthy Volunteers," *Archives of General Psychiatry* 51, no. 1 (1994): 20–27.

16. Mitchell E. Berman and Emil F. Coccaro, "Neurobiologic Correlates of Violence: Relevance to Criminal Responsibility," *Behavioral Sciences and the Law* 16 (1998): 303–18.

17. Michael J. Raleigh et al., "Serotonergic Mechanisms Promote Dominance Acquisition in Adult Male Vervet Monkeys," *Brain Research* 559 (1991): 181–90.

18. Rhona Limson et al., "Personality and Cerebrospinal Fluid Monoamine Metabolites in Alcoholics and Controls," *Archives of General Psychiatry* 48, no. 5 (1991): 437–41.

19. Matti Virkkunen, A. Nuutila, F. K. Goodwin, and M. Linnoila, "Cerebrospinal Fluid Monoamine Metabolite Levels in Male Arsonists," *Archives of General Psychiatry* 44, no. 3 (1987): 241–47.

20. Gary B. Melton et al. *Psychological Evaluations for the Courts: A Handbook for Mental Health Professionals and Lawyers* (New York: Guilford Press, 1997), p. 191.

21. Ibid., pp. 191–93.

22. Gyorgy Csaba and Kornélia Tekes, "Is the Brain Hormonally Imprintable?" *Brain & Development* 27 (2005): 465–71.

23. Csaba Gyorgy et al., "Effect of Neonatal-Endorphin Imprinting on Sexual Behavior and Brain Serotonin Level in Adult Rats," *Life Sciences* 73 (2003): 103–14.

24. Csaba Gyorgy et al., "Effect of Mianserin Treatment at Weaning with the Serotonin Antagonist Mianserin on the Brain Serotonin and Cerebrospinal Fluid Nocistatin Level of Adult Female Rats: A Case of Late Imprinting," *Life Sciences* 75 (2004): 939–46.

25. Ilona Vathy et al., "Modulation of Catecholamine Turnover Rate in Brain Regions of Rats Exposed Prenatally to Morphine," *Brain Research* 662 (1994): 209–15.

26. Hoau-Yan Wang et al., "Prenatal Cocaine Exposure Selectively Reduces Mesocortical Dopamine Release," *Journal of Pharmacology and Experimental Therapeutics* 273 (1995): 121–25.

27. David R. Owen, "The 47, XYY Male: A Review," *Psychological Bulletin* 78, no. 3 (1972): 209–33.

28. Jan Volavka, *The Neurobiology of Violence* (Washington, DC: American Psychiatric Press, 1995), p. 70.

29. Owen, "The 47, XYY Male."

30. Herman A Witkin et al., "Criminality in XYY and XXY Men," *Science* 193, no. 4253 (1976): 547–55.

31. Bernard Bioulac et al., "Biogenic Amines in 47, XYY Syndrome," *Neuropsychobiology* 4, no. 6 (1978): 366–70; and Volavka, *Neurobiology of Violence*, p. 51.

32. Several studies, such as Kandel et al., "IQ as a Protective Factor for Subjects at High Risk for Antisocial Behavior," *Journal of Consulting and Clinical Psychology* 56, no. 2 (1988): 224–26, have shown that intelligent people are less likely to engage in criminal acts.

33. Volavka, *Neurobiology of Violence*, pp. 124–25.

34. Jan Volavka, "The Neurobiology of Violence: An Update," *Journal of Neuropsychiatry and Clinical Neurosciences* 11 (1999): 307–14.

35. John H. Morton et al., "A Clinical Study of Premenstrual Tension," *American Journal of Obstetrics and Gynecology* 65, no. 6 (1953): 1182–91; P. T. d'Orbán and James Dalton, "Violent Crime and the Menstrual Cycle," *Psychological Medicine* 10, no. 2 (1980): 353–59; and Volavka, *Neurobiology of Violence*, pp. 74–76.

36. "Premenstrual Syndrome (PMS)," October 27, 2006, http://www.mayo clinic.com/health/premenstrual-syndrome/DS00134.

37. Don R. Cherek et al., "Effects of Chronic Paroxetine Administration on Measures of Aggressive and Impulsive Responses of Adult Males with a History of Conduct Disorder," *Psychopharmacology* 159 (2002): 266–74; and Alyson J. Bond, "Antidepressant Treatments and Human Aggression," *European Journal of Pharmacology* 526, no. 1–3 (2005): 218–25.

38. Steven Rose, *The Future of the Brain: The Promise and Perils of Tomorrow's Neuroscience* (Oxford: Oxford University Press, 2005), p. 271.

39. Amanda C. Pustilnik, "Violence on the Brain: A Critique of Neuroscience in Criminal Law," Harvard Law School Faculty Scholarship Series, paper 14, 2008, http://lsr.nellco.org/harvard_faculty/14; Rose 2005, pp. 276–77.

CHAPTER 13: MORALITY'S END

1. Harry J. Maihafer, *Brave Decisions: Fifteen Profiles in Courage and Character from American Military History* (Dulles, VA: Brassey's, 1995), p. 201.

2. Ibid., pp. 196–210.

3. Michael S. Gazzaniga, *The Ethical Brain* (New York: Dana Press, 2005), p. 148.

4. Ibid., pp. 148–49.

5. Vilayanur S. Ramachandran, "Anosognosia in Parietal Lobe Syndrome," in *Essential Sources in the Scientific Study of Consciousness*, ed. Bernard J. Baars et al. (Cambridge, MA: MIT Press, 2003), pp. 805–30.

6. Ibid.

7. Michael S. Gazzaniga, "Cerebral Specialization and Interhemispheric Communication: Does the Corpus Collosum Enable the Human Condition?" *Brain* 123 (2000): 1293–1326.

8. Gazzaniga, *The Ethical Brain*, pp. 149–50.

9. Ibid., pp. 156–57.

10. Dennis R. Hill and Michael A. Persinger, "Application of Transcerebral, Weak (1 microT) Complex Magnetic Fields and Mystical Experiences: Are They Generated by Field-Induced Dimethyltryptamine Release from the Pineal Organ?" *Perceptual and Motor Skills* 97 (2003): 1049–50.

11. Gazzaniga, *The Ethical Brain*, p. 161.

12. Ibid., p. 90.

CHAPTER 14: THE DEPTHS OF CONSCIOUSNESS

1. The person who worked with Bauby was actually a representative of the publishing company.

2. Thomas Mallon, "In the Blink of an Eye," *New York Times*, June 15, 1997.

3. Jean D. Bauby, *The Diving Bell and the Butterfly: A Memoir of Life in Death* (New York: Vintage International, 1997), pp. 3–5.

4. J. Fodor, *A Theory of Content and Other Essays* (Cambridge, MA: MIT Press), p. 196.

5. The comparison of the moral agent to an F-16 pilot is not meant to suggest the agent is some kind of dualistic homunculus that controls the brain. The analogy is meant only to reveal the fallacy in the argument from brain disorder.

6. Of course, these actions have not actually been shown to be determined, despite what some scientists may think. As we have discussed earlier, the results of those experiments are open to many interpretations.

7. On a separate note, we should not forget the discussions earlier in the book regarding the weaknesses of these studies in addressing the free will issue.

8. David L. Rosenhan, "On Being Sane in Insane Places," *Science* 179 (1973): 250–58; and James Rachels, *The Elements of Moral Philosophy* (Philadelphia: Temple University Press, 1986), pp. 62–63.

CHAPTER 15: A CHALLENGE FOR EXPERIENCE

1. In the original story, from James M. Olson, *Fair Play: The Moral Dilemmas of Spying* (Washington, DC: Potomac Books, 2006), he is code-named ZTAC-CORDEON.

2. This scenario was taken from Olson, *Fair Play*, pp. 105–109.

3. Peter C. Wason, "Reasoning," in *New Horizons in Psychology*, ed. Brian M. Foss (Harmondsworth, UK: Penguin Books, 1966).

4. Richard A. Griggs and Jerome R. Cox, "The Elusive Thematics Material Effect in Wason's Selection Task," *British Journal of Psychology* 73 (1982): 407–20; and Peter C. Wason and Philip N. Johnson-Laird, *Psychology of Reasoning: Structure and Content* (London: Batsford, 1972).

5. A version of this premise has been mathematically proven by Gödel's incompleteness theorem. This famous proof shows that there are certain problems that every formal system, no matter how complex, could in principle never solve.

6. In developing this and the following chapters of this book, I have been strongly influenced by the work of philosophers Hubert Dreyfus and John Searle.

7. This example is an adapted form of one suggested by Larry Wright in his paper "Argument and Deliberation: A Plea for Understanding," *Journal of Philosophy* 92, no. 11 (1995): 565–85.

CHAPTER 16: THE BOUNDLESSNESS OF REASON

1. Robert Frost, "Stopping by Woods on a Snowy Evening," in *The Poetry of Robert Frost*, ed. Edward Connery Lathem. Copyright 1923, © 1969 by Henry Holt and Company, Inc., renewed 1951, by Robert Frost. Reprinted with permission of Henry Holt and Company.

CHAPTER 17: RISE OF THE MORAL AGENT

1. Patrick Hughes, "The Meteorologist Who Started a Revolution," *Weatherwise*, January 1998.

2. Alfred Wegener, *The Origin of Continents and Oceans* (New York: Dover, 1966), pp. 5–23.

3. Hughes, "The Meteorologist Who Started a Revolution."

4. For more on these topics, see Thomas S. Kuhn, *The Structure of Scientific Revolutions*, 2nd ed. (Chicago: University of Chicago Press, 1996).

5. In constructing the relation this way, I rely on Russ Shafer-Landau and Joel Feinberg, *Reason and Responsibility: Readings in Some Basic Problems of Philosophy* (Belmont, CA: Wadsworth, 2004), pp. 387–88.

6. Quoted in Jean Burns, "Does Consciousness Perform a Function Independently of the Brain?" *Frontier Perspectives* 2, no. 1 (1991): 19–34.

CHAPTER 18: THE PALACE OF THE MIND

1. Based on a passage from Tullius Cicero, *De Oratore*, trans. E. W. Button and H. Rackham (Cambridge, MA: Loeb–Harvard University Press, 1976).

2. Quintilian, *Institutio Oratoria*, vol. 4., trans. H. E. Butler (New York: Loeb Classics Library, 1936).

3. Jonathan D. Spence, *The Memory Palace of Matteo Ricci* (New York: Viking, 1984), pp. 266–68.

4. In fact, US law takes this issue into account in the concept of felony murder. In the course of an armed robbery, even an accidental killing will result in a charge of murder in the first degree.

5. Of course, this is not to imply that evolution is any kind of intentionally progressive system. I'm only saying that consciousness is a powerful ability.

6. Fyodor Dostoevsky, *Notes from Underground*, trans. Ralph E. Matlaw (New York: E. P. Dutton and Co., 1960), p. 31.

7. Ibid., pp. 9–10.

BIBLIOGRAPHY

Ackerman, Sandra. *Discovering the Brain.* Washington, DC: National Academy Press, 1992.

Adolphs, Ralph. "Social Cognition and the Human Brain." *Trends in Cognitive Sciences* 3, no. 12 (1999): 469–79.

Apollodorus, *The Library of Greek Mythology,* translated by Robin Hard. Oxford: Oxford University Press, 1999.

Atkins, Kim. *Self and Subjectivity.* Malden, MA: Blackwell, 2005.

Baars, Bernard J. *In the Theater of Consciousness: The Workspace of the Mind.* New York: Oxford University Press, 1997.

Baars, Bernard J., et al. *Essential Sources in the Scientific Study of Consciousness.* Cambridge, MA: MIT Press, 2003.

Barad, Mark, et al. "Rolipram, a Type IV-Specific Phosphodiesterase Inhibitor, Facilitates the Establishment of Long-Lasting Long-Term Potentiation and Improves Memory." *Proceedings of the National Academy of Sciences* 95 (1998): 15020–25.

Bauby, Jean D. *The Diving Bell and the Butterfly: A Memoir of Life in Death.* New York: Vintage International, 1997.

Bechara, A., A. R. Damasio, H. Damasio, and S. W. Anderson. "Insensitivity to Future Consequences Following Damage to Human Prefrontal Cortex." *Cognition* 50 (1994): 7–15.

Bechara, A., H. Damasio, and A. R. Damasio. "Emotion, Decision Making and the Orbitofrontal Cortex." *Cerebral Cortex* 10 (2000): 295–307.

Berman, Mitchell E., and Emil F. Coccaro. "Neurobiologic Correlates of Violence: Relevance to Criminal Responsibility." *Behavioral Sciences and the Law* 16 (1998): 303–18.

Bioulac, Bernard, et al. "Biogenic Amines in 47, XYY Syndrome." *Neuropsychobiology* 4, no. 6 (1978): 366–70.

Birbaumer, Niels. "Effect of Regional Anesthesia on Phantom Limb Pain Are Mirrored in Changes in Cortical Reorganization." *Journal of Neuroscience* 17, no. 14 (1997): 5503–5508.

Blank, Robert H. *Brain Policy: How the New Neuroscience Will Change Our Lives and Our Politics.* Washington, DC: Georgetown University Press, 1999.

Bolbecker, Amanda R. "Two Asymmetries Governing Neural and Mental Timing." *Consciousness and Cognition* 11 (2002): 265–72.

Bond, Alyson J. "Antidepressant Treatments and Human Aggression." *European Journal of Pharmacology* 526, no. 1–3 (2005): 218–25.

Brandt, Richard. "A Moral Principle about Killing." In Marvin Kohl, *Beneficent Euthanasia.* Amherst, NY: Prometheus Books, 1972.

Breitmeyer, Bronu G. "In Support of Pockett's Critique of Libet's Studies of the Time Course of Consciousness." *Consciousness and Cognition* 11 (2002): 280–83.

Burns, Jean. "Does Consciousness Perform a Function Independently of the Brain?" *Frontier Perspectives* 2, no. 1 (1991): 19–34.

Caldwell, John A., Jr., et al. "A Double-Blind, Placebo-Controlled Investigation of the Efficacy of Modafinil for Sustaining the Alertness and Performance of Aviators: A Helicopter Simulator Study." *Psychopharmacology* 150, no. 3 (2000): 272–82.

Calvin, William H. *A Brief History of the Mind.* Oxford: Oxford University Press, 2004.

Casebeer, William D. "Moral Cognition and Its Neural Constituents." *Nature Reviews Neuroscience* 4 (2003): 841–46.

Cavedini, Paolo, et al. "Frontal Lobe Dysfunction in Obsessive-Compulsive Disorder and Major Depression: A Clinical-Neuropsychological Study." *Psychiatry Research* 78, no. 1–2 (1998): 21–28.

Chalmers, David J. *The Conscious Mind: In Search of a Fundamental Theory.* Oxford: Oxford University Press, 1996.

———. "Facing Up to the Problem of Consciousness." *Journal of Consciousness Studies* 2, no. 3 (1995): 200–19.

———. *Philosophy of Mind: Classical and Contemporary Readings.* Oxford: Oxford University Press, 2002.

Chappell, T. D. J. *Understanding Human Goods: A Theory of Ethics.* Edinburgh: Edinburgh University Press, 1998.

Chatterjee, Anjan. "Cosmetic Neurology: The Controversy over Enhancing Movement, Mentation, and Mood." *Neurology* 63 (2004): 968–74.

Cherek, Don R., et al. "Effects of Chronic Paroxetine Administration on Measures of Aggressive and Impulsive Responses of Adult Males with a History of Conduct Disorder." *Psychopharmacology* 159 (2002): 266–74.

Churchland, Patricia S. "Self-Representation in Neural Systems." *Annals of the New York Academy of Sciences* 1001 (2003): 31–38.

Cicero, Tullius. *De Oratore*. Translated by E. W. Button and H. Rackham. Cambridge, MA: Loeb–Harvard University Press, 1976.

Conant, James B. *The Overthrow of the Phlogiston Theory: The Chemical Revolution of 1775–1789*. Cambridge, MA: Harvard University Press, 1956.

Crick, Francis. *The Astonishing Hypothesis: The Scientific Search for the Soul*. New York: Charles Scribner's Sons, 1994.

Crick, Francis, and Christof Koch. "Consciousness and Neuroscience." *Cerebral Cortex* 8 (1998): 97–107.

Csaba, Gyorgy, et al. "Effect of Mianserin Treatment at Weaning with the Serotonin Antagonist Mianserin on the Brain Serotonin and Cerebrospinal Fluid Nocistatin Level of Adult Female Rats: A Case of Late Imprinting." *Life Sciences* 75 (2004): 939–46.

———. "Effect of Neonatal β-Endorphin Imprinting on Sexual Behavior and Brain Serotonin Level in Adult Rats." *Life Sciences* 73 (2003): 103–14.

Csaba, Gyorgy, and Kornélia Tekes. "Is the Brain Hormonally Imprintable?" *Brain & Development* 27 (2005): 465–71.

Damasio, Antonio R. *Descartes' Error: Emotion, Reason and the Human Brain*. New York: Avon Books, 1994.

———. *The Feeling of What Happens: Body and Emotion in the Making of Consciousness*. San Diego, CA: Harvest Books, 2000.

———. *Looking for Spinoza*. New York: Vintage Books, 2004.

Darrow, Clarence. *Crime and Criminals: An Address Delivered to the Prisoners in the Chicago County Jail*. Chicago: Charles H. Kerr and Company, 1919.

———. "A Plea for Mercy." In Ashley H. Thorndike, *Modern Eloquence*, vol. 6, 80–85. New York: P. F. Collier & Son, 1936.

Den Boer, Johan A. "Social Anxiety Disorder/Social Phobia: Epidemiology, Diagnosis, Neurobiology, and Treatment." *Comprehensive Psychiatry* 46, no. 6 (2000): 405–15.

Dennett, Daniel C. *Breaking the Spell: Religion as a Natural Phenomenon*. New York: Viking, 2006.

———. *Consciousness Explained*. Boston, MA: Little, Brown and Company, 1991.

———. *Elbow Room: The Varieties of Free Will Worth Wanting*. Cambridge, MA: MIT Press, 1984.

————. *Freedom Evolves.* New York: Viking, 2003.

————. "The Self as a Responding—and Responsible—Artifact." *Annals of the New York Academy of Sciences* 1001 (2003): 39–50.

Descartes, René. *Meditations on First Philosophy.* Translated by John Veitch in 1901. http://www.wright.edu/cola/descartes/.

Doidge, Norman. *The Brain That Changes Itself: Stories of Personal Triumph from the Frontiers of Brain Science.* New York: Viking, 2007.

Dolan, Raymond J. "On the Neurology of Morals." *Nature Neuroscience* 2, no. 11 (1999): 927–29.

d'Orbán, P. T., and James Dalton. "Violent Crime and the Menstrual Cycle." *Psychological Medicine* 10, no. 2 (1980): 353–59.

Dostoevsky, Fyodor. *Notes from Underground.* Translated by Ralph E. Matlaw. New York: E. P. Dutton and Co., 1960.

Dreyfus, Hubert L. *What Computers Still Can't Do: A Critique of Artificial Reason.* Cambridge, MA: MIT Press, 1992.

Earleywine, Mitch. *Mind-Altering Drugs: The Science of Subjective Experience.* Oxford: Oxford University Press, 2005.

Eccles, John C. *How the Self Controls Its Brain.* New York: Springer-Verlag, 1994.

Edelman, Gerald M. *Bright Air, Brilliant Fire: On the Matter of the Mind.* New York: Basic Books, 1992.

————. *The Remembered Present: A Biological Theory of Consciousness.* New York: Basic Books, 1989.

————. *A Universe of Consciousness: How Matter Becomes Imagination.* New York: Basic Books, 2000.

————. *Wider Than the Sky: The Phenomenal Gift of Consciousness.* New Haven, CT: Yale University Press, 2005.

"Expanding *Nature Neuroscience.*" *Nature Neuroscience* 8 (2005): 1.

Fodor, J. *A Theory of Content and Other Essays.* Cambridge, MA: MIT Press, 1990.

Franzén, Torkel. *Gödel's Theorem: An Incomplete Guide to Its Use and Abuse.* Wellesley, MA: A. K. Peters, 2005.

French, Peter A., et al. *Free Will and Moral Responsibility.* Boston: Blackwell, 2005.

Garland, Brent. *Neuroscience and the Law: Brain, Mind and the Scales of Justice.* New York: Dana Press, 2004.

Gazzaniga, Michael S. "Cerebral Specialization and Interhemispheric Communication: Does the Corpus Collosum Enable the Human Condition?" *Brain* 123 (2000): 1293–1326.

————. *The Ethical Brain.* New York: Dana Press, 2005.

————, ed. *The New Cognitive Neurosciences.* Cambridge, MA: MIT Press, 2000.

Gazzaniga, Michael S., et al. *Cognitive Neuroscience: The Biology of the Mind.* New York: W. W. Norton, 2002.

Georgopoulos, Apostolos P. "Neural Mechanisms of Motor Cognitive Processes: Functional MRI and Neurophysiological Studies." In *The New Cognitive Neurosciences*, edited by Michael S. Gazzaniga, 525–38. Cambridge, MA: MIT Press, 2000.

Glannon, Walter. *The Mental Basis of Responsibility.* Aldershot, UK: Ashgate, 2002.

Goetz, Stewart. "Frankfurt-Style Counterexamples and Begging the Question." *Midwest Studies in Philosophy* 29 (2005): 83–105.

Goldberg, Elkhonon. *The Executive Brain: The Frontal Lobes and the Civilized Mind.* Oxford: Oxford University Press, 2001.

Gomes, Gilberto. "On Experimental and Philosophical Investigations of Mental Timing: A Response to Commentary." *Consciousness and Cognition* 11 (2002): 304–307.

———. "Problems in the Timing of Conscious Experience." *Consciousness and Cognition* 11 (2002): 191–97.

Goswami, Amit. *The Physicists' View of Nature: The Quantum Revolution.* New York: Springer, 1992.

Greene, Joshua. "From Neural 'Is' to Moral 'Ought': What Are the Moral Implications of Neuroscientific Moral Psychology?" *Nature Reviews Neuroscience* 4 (2003): 847–50.

Griggs, Richard A., and Jerome R. Cox. "The Elusive Thematics Material Effect in Wason's Selection Task." *British Journal of Psychology* 73 (1982): 407–20.

Haggard, Patrick, and Martin Eimer. "On the Relation between Brain Potentials and the Awareness of Voluntary Movements." *Experimental Brain Research* 126 (1999): 128–33.

Hardcastle, Valerie G. "The Elusive Illusion of Sensation." *Behavioral and Brain Sciences* 27, no. 5 (2004): 662–63.

"He Listens to the Brain's 'Sur.'" May 29, 2001. http://www.rediff.com/news/may401us.htm.

Herodotus. *Clio.* http://www.greektexts.com/library/Herodotus/Clio/eng/329.html.

Higdon, Hal. *Leopold and Loeb: The Crime of the Century.* Champaign: University of Illinois Press, 1999.

Hill, Dennis R., and Michael A. Persinger. "Application of Transcerebral, Weak (1 microT) Complex Magnetic Fields and Mystical Experiences: Are They Generated by Field-Induced Dimethyltryptamine Release from the Pineal Organ?" *Perceptual and Motor Skills* 97 (2003): 1049–50.

Hinde, Robert A. *Why Good Is Good: The Sources of Morality.* New York: Routledge, 2002.

Hofstadter, Douglas. *Gödel, Escher, Bach: An Eternal Golden Braid.* New York: Basic Books, 1979.

————. *I Am a Strange Loop*. New York: Basic Books, 2007.

Honderich, Ted. *How Free Are You? The Determinism Problem*. Oxford: Oxford University Press, 2002.

Howard-Snyder, Daniel, and Jeff Jordan. *Faith, Freedom and Rationality*. Lanham, MD: Rowman and Littlefield, 1996.

Hubbard, Timothy L., and Jamshed J. Barucha. "Judged Displacement in Apparent Vertical and Horizontal Motion." *Perception and Psychophysics* 44, no. 3 (1988): 211–21.

Hughes, Patrick. "The Meteorologist Who Started a Revolution." *Weatherwise*, January 1998.

Hugo, Victor. *Les Misérables*. Translated by Lee Fahnestock and Norman MacAfee. New York: Signet Classics, 1987.

Huxley, Thomas H. "On the Hypothesis That Animals Are Automata, and Its History." In *Collected Essays* by T. H. Huxley. Boston, MA: Adamant Media Corporation, 2000 [1874].

Johnson, Steven. *Mind Wide Open: Your Brain and the Neuroscience of Everyday Life*. New York: Scribner, 2004.

Johnson-Laird, Philip N., et al. "Reasoning and a Sense of Reality." *British Journal of Psychology* 63 (1972): 392–400.

Jones, Lynette A. "Motor Illusions: What Do They Reveal about Proprioception?" *Psychological Bulletin* 103 (1998): 72–86.

Joordens, Steve, et al. "When Timing the Mind One Also Should Mind the Timing: Biases in the Measurement of Voluntary Actions." *Consciousness and Cognition* 11 (2002): 231–40.

Joyce, Richard. *The Myth of Morality*. Cambridge: Cambridge University Press, 2001.

Kahn, Charles H. *The Art and Thought of Heraclitus: An Edition of the Fragments with Translation and Commentary*. Cambridge: Cambridge University Press, 1979.

Kalian, Moshe, et al. "Political Assassins—The Psychiatric Perspective and Beyond." *Medicine and Law* 22, no. 1 (2003): 113–30.

Kandel, Elizabeth, et al. "IQ as a Protective Factor for Subjects at High Risk for Antisocial Behavior." *Journal of Consulting and Clinical Psychology* 56, no. 2 (1988): 224–26.

Kane, Robert. *The Oxford Handbook of Free Will*. Oxford: Oxford University Press, 2002.

Klein, Stanley. "Libet's Research on the Timing of Conscious Intention to Act: A Commentary." *Consciousness and Cognition* 11 (2002): 273–79.

————. "Libet's Research on the Timing of Mental Events: A Commentary on the Commentaries." *Consciousness and Cognition* 11 (2002): 326–33.

Koch, Christof. *The Quest for Consciousness: A Neurobiological Approach*. Englewood, CO: Roberts and Company, 2004.

Kohl, Marvin. *Beneficent Euthanasia*. Amherst, NY: Prometheus Books, 1972.

Kramer, Peter D. *Listening to Prozac: The Landmark Book about Antidepressants and the Remaking of the Self.* New York: Penguin Books, 1997.

Kuhn, Thomas S. *The Structure of Scientific Revolutions*, 2nd ed. Chicago: University of Chicago Press, 1996.

Laplace, Pierre. *A Philosophical Essay on Probabilities*. Translated by F. W. Truscott and F. L. Emory. New York: Dover, 1951.

Lau, Hakwan C., et al. "On Measuring the Perceived Onset of Spontaneous Actions." *Journal of Neuroscience* 26, no. 27 (2006): 7265–71.

LeDoux, Joseph. "The Self: Clues from the Brain." *Annals of the New York Academy of Sciences* 1001 (2003): 295–304.

———. *Synaptic Self: How Our Brains Become Who We Are*. New York: Penguin Books, 2002.

Levere, Trevor H. *Transforming Matter: A History of Chemistry from Alchemy to the Buckyball*. Baltimore: Johns Hopkins University Press, 2001.

Lhermitte, François. "Human Autonomy and the Frontal Lobes. Part I: Imitation and Utilization Behavior: A Neuropsychological Study of 75 Patients." *Annals of Neurology* 19, no. 4 (1986): 326–34.

———. "'Utilization Behavior' and Its Relation to Lesions of the Frontal Lobes." *Brain* 106, no. 2 (1983): 237–55.

Libet, Benjamin. "Do We Have Free Will?" In *Volitional Brain*, edited by Benjamin Libet et al. Thoverton, UK: Imprint Academic, 1999.

———. *Mind Time: The Temporal Factor in Consciousness*. Cambridge, MA: Harvard University Press, 2004.

———. "The Timing of Mental Events: Libet's Experimental Findings and Their Implications." *Consciousness and Cognition* 11 (2002): 291–99.

Libet, Benjamin, et al., eds. *The Volitional Brain: Toward a Neuroscience of Free Will*. Thoverton, UK: Imprint Academic, 1999.

Lim, Gerald T., et al. "Clinicopathologic Case Report: Akinetic Mutism with Findings of White Matter Hyperintensity." *Journal of Neuropsychiatry and Clinical Neurosciences* 14 (2002): 214–21.

Limson, Rhona, et al. "Personality and Cerebrospinal Fluid Monoamine Metabolites in Alcoholics and Controls." *Archives of General Psychiatry* 48, no. 5 (1991): 437–41.

Linnoila, Markku, et al. "Low Cerebrospinal Fluid 5-Hydroxyindoleacetic Acid Concentration Differentiates Impulsive from Nonimpulsive Violent Behavior." *Life Sciences* 33 (1983): 2609–14.

Locke, John. *An Essay Concerning Human Understanding*. Edited by A. C. Fraser. New York: Dover, 1959 [1689].

Maihafer, Harry J. *Brave Decisions: Fifteen Profiles in Courage and Character from American Military History*. Dulles, VA: Brassey's, 1995.

Mallon, Thomas. "In the Blink of an Eye." *New York Times*, June 15, 1997.

Marcus, Steven J. *Neuroethics: Mapping the Field.* New York: Dana Press, 2002.

Martin, Kevan. "Time Waits for No Man." *Nature* 429, no. 20 (2004): 243–44.

Martin, Mike W. *From Morality to Mental Health: Virtue and Vice in a Therapeutic Culture.* Oxford: Oxford University Press, 2006.

Masters, Roger D., and Michael T. McGuire. *The Neurotransmitter Revolution: Serotonin, Social Behavior, and the Law.* Carbondale: Southern Illinois University Press, 1994.

McCabe Sean E., et al. "Non-medical Use of Prescription Stimulants among US College Students: Prevalence and Correlates from a National Survey." *Addiction* 99 (2005): 96–106.

Mele, Alfred R. *Autonomous Agents: From Self-Control to Autonomy.* Oxford: Oxford University Press, 1995.

Melton, Gary B., et al. *Psychological Evaluations for the Courts: A Handbook for Mental Health Professionals and Lawyers.* New York: Guilford Press, 1997.

"Messing with Our Minds." *Independent*, January 18, 2005.

Metzinger, Thomas. *Neural Correlates of Consciousness: Empirical and Conceptual Questions.* Cambridge, MA: MIT Press, 2000.

Milan, Wil. "Fear and Awe: Eclipses through the Ages." January 18, 2000. http://www.space.com/scienceastronomy/solarsystem/lunar_lore_000118.html.

Miller, Richard B. *Casuistry and Modern Ethics: A Poetics of Practical Reasoning.* Chicago: University of Chicago Press, 1996.

Moll, Jorge, R. de Oliveira-Souza, and P. J. Eslinger. "Morals and the Human Brain: A Working Model." *Neuroreport* 14, no. 3 (2003): 299–305.

Moll, Jorge, et al. "The Neural Basis of Human Moral Cognition." *Nature Reviews Neuroscience* 6 (2005): 799–809.

Moreno, Jonathan D. *Mind Wars: Brain Research and National Defense.* Washington, DC: Dana Press, 2006.

Morton, John H., et al. "A Clinical Study of Premenstrual Tension." *American Journal of Obstetrics and Gynecology* 65, no. 6 (1953): 1182–91.

Nahmias, Eddy. "When Consciousness Matters: A Critical Review of Daniel Wegner's *The Illusion of Conscious Will*." *Philosophical Psychology* 15, no. 4 (2002): 527–42.

"Narcolepsy More Common in Men, Often Originates in Their 20s." http://www.mayoclinic.org/news2002-rst/986.html.

Nelkin, Dorothy, and M. Susan Lindee. *The DNA Mystique: The Gene as a Cultural Icon.* New York: W. H. Freeman, 1995.

Nelson, Randy J. *Biology of Aggression.* Oxford: Oxford University Press, 2006.

Obhi, Sukhvinder S., and Patrick Haggard. "Free Will and Free Won't." *American Scientist* 92 (2004): 358–65.

O'Connor, Daniel J. *Free Will.* London: Macmillan, 1971.

Olson, James M. *Fair Play: The Moral Dilemmas of Spying.* Washington, DC: Potomac Books, 2006.

Overbye, Dennis. "Free Will: Now You Have It, Now You Don't." *New York Times,* January 2, 2007.

Owen, David R. "The 47, XYY Male: A Review." *Psychological Bulletin* 78, no. 3 (1972): 209–33.

Penfield, Wilder. *The Mystery of Mind.* Princeton, NJ: Princeton University Press, 1975.

Penrose, Roger. *The Emperor's New Mind.* Oxford: Oxford University Press, 1989.

Pereboom, Derk. *Living without Free Will.* Cambridge: Cambridge University Press, 2001.

Pink, Thomas. *The Psychology of Freedom.* Cambridge: Cambridge University Press, 1996.

Pinker, Steven. *The Blank Slate: The Modern Denial of Human Nature.* New York: Viking, 2002.

Plato. *Republic.* Translated by G. M. A. Grube. Indianapolis: Hacket, 1992.

"Premenstrual Syndrome (PMS)." October 27, 2006. http://www.mayoclinic.com/health/premenstrual-syndrome/DS00134.

Price-Huish, Cecille. "Born to Kill? Aggression Genes and Their Potential Impact on Sentencing and the Criminal Justice System." *Southern Methodist University Law Review* (1997): 603–10.

Purdy, Michael. "Researchers Use Brain Scans to Predict Behavior." Washington School of Medicine in St. Lewis Online, November 29, 2005. http://mednews.wustl.edu/news/page/normal/6248.html.

Pustilnik, Amanda C. "Violence on the Brain: A Critique of Neuroscience in Criminal Law." Harvard Law School Faculty Scholarship Series. Paper 14, 2008. http://lsr.nellco.org/harvard_faculty/14.

Quintilian. *Institutio Oratoria,* vol. 4. Translated by H. E. Butler. New York: Loeb Classics Library, 1936.

Rachels, James. *The Elements of Moral Philosophy.* Philadelphia: Temple University Press, 1986.

Raine, Adrian. *The Psychopathology of Crime: Criminal Behavior as a Clinical Disorder.* San Diego, CA: Academic Press, 1993.

Raleigh, Michael J., et al. "Serotonergic Mechanisms Promote Dominance Acquisition in Adult Male Vervet Monkeys." *Brain Research* 559 (1991): 181–90.

Ramachandran, Vilayanur S. "Anosognosia in Parietal Lobe Syndrome." In *Essential Sources in the Scientific Study of Consciousness,* edited by Bernard J. Baars et al., 805–30. Cambridge, MA: MIT Press, 2003.

———. Quoted in "The Zombie Within." *New Scientist,* September 5, 1998, p. 35.

Restak, Richard. *The Brain Has a Mind of Its Own: Insights from a Practicing Neurologist.* New York: Three Rivers Press, 1993.

———. *The Naked Brain: How the Emerging Neurosociety Is Changing How We Live, Work, and Love.* Easton, PA: Harmony Press, 2006.

———. *The New Brain: How the Modern Age Is Rewiring Your Mind.* New York: Rodale, 2003.

Rose, Steven. *The Future of the Brain: The Promise and Perils of Tomorrow's Neuroscience.* Oxford: Oxford University Press, 2005.

Rosenhan, David L. "On Being Sane in Insane Places." *Science* 179 (1973): 250–58.

Ryle, Gilbert. *The Concept of Mind.* London: Hutchinson & Company, 1949.

Sapir, Ayelet, et al., "Brain Signals for Spatial Attention Predict Performance in a Motion Discrimination Task." *Proceedings of the National Academy of Sciences* 103, no. 49 (2005): 17810–15.

Satinover, Jeffrey. *The Quantum Brain: The Search for Freedom and the Next Generation of Man.* New York: John Wiley and Sons, 2001.

Scott, George P. *Atoms of the Living Flame: An Odyssey into Ethics and the Physical Chemistry of Free Will.* Lanham, MD: University Press of America, 1985.

Searle, John R. *Mind: A Brief Introduction.* Oxford: Oxford University Press, 2004.

———. "Is the Brain's Mind a Computer Program?" *Scientific American*, October 1990.

Shafer-Landau, Russ, and Joel Feinberg. *Reason and Responsibility: Readings in Some Basic Problems of Philosophy.* Belmont, CA: Wadsworth, 2004.

"Solar Eclipses in History and Mythology: Historical Observations of Solar Eclipses." Bibliotheca Alexandria Online. March 29, 2006. http://www.bibalex.org/eclipse2006/HistoricalObservationsofSolarEclipses.htm.

Spence, Jonathan D. *The Memory Palace of Matteo Ricci.* New York: Viking, 1984.

Spence, Sean A., and Chris D. Frith. "Towards a Functional Anatomy of Volition." *Journal of Conscious Studies* 6 (1999): 11–29.

Spence, Sean A., et al. "A PET Study of Voluntary Movement in Schizophrenic Patients Experiencing Passivity Phenomena." *Brain* 120, no. 11 (1997): 1997–2011.

Sternberg, Eliezer J. *Are You a Machine? The Brain, the Mind, and What It Means to Be Human.* Amherst, NY: Humanity Books, 2007.

Stump, Eleonore. "Libertarian Freedom and the Principle of Alternative Possibilities." In *Faith, Freedom and Rationality*, edited by Daniel Howard-Snyder and Jeff Jordan, 73–88. Lanham, MD: Rowman and Littlefield, 1996.

Stuss, Donald T., et al. "The Frontal Lobes Are Necessary for 'Theory of Mind.'" *Brain* 124 (2001): 279–86.

Tanaka, Yutaka, et al. "Forced Hyperphasia and Environmental Dependency Syndrome." *Journal of Neurology, Neurosurgery and Psychiatry* 68, no. 2 (2000): 224–26.

Taylor, Maxwell D. *The Uncertain Trumpet*. New York: Harper & Brothers, 1960.

Taylor, Stuart, Jr. "CAT Scans Said to Show Shrunken Hinckley Brain." *New York Times*, June 2, 1982.

Thorndike, Ashley H. *Modern Eloquence*. New York: P. F. Collier & Son, 1936.

Trevena, Judy A., and Jeff Miller. "Cortical Movement Preparation Before and After a Conscious Decision to Move." *Consciousness and Cognition* 11, no. 2 (2002): 162–90.

Turk, David J., et al. "Out of Contact, Out of Mind: The Distributed Nature of the Self." *Annals of the New York Academy of Sciences* 1001 (2003): 65–78.

Vathy, Ilona, et al. "Modulation of Catecholamine Turnover Rate in Brain Regions of Rats Exposed Prenatally to Morphine." *Brain Research* 662 (1994): 209–15.

Virkkunen, Matti, A. Nuutila, F. K. Goodwin, and M. Linnoila. "Cerebrospinal Fluid Monoamine Metabolite Levels in Male Arsonists." *Archives of General Psychiatry* 44, no. 3 (1987): 241–47.

Virkkunen, Matti, et al. "CSF Biochemistries, Glucose Metabolism, and Diurnal Activity Rhythms in Alcoholic, Violent Offenders, Fire Setters, and Healthy Volunteers." *Archives of General Psychiatry* 51, no. 1 (1994): 20–27.

Volavka, Jan. *The Neurobiology of Violence*. Washington, DC: American Psychiatric Press, 1995.

———. "The Neurobiology of Violence: An Update." *Journal of Neuropsychiatry and Clinical Neurosciences* 11 (1999): 307–14.

Walter, Henrik. "Neurophilosophy of Free Will." In *The Oxford Handbook of Free Will*, edited by Robert Kane, 565–76. Oxford: Oxford University Press, 2002.

Wang, Hoau-Yan, et al. "Prenatal Cocaine Exposure Selectively Reduces Meso-cortical Dopamine Release." *Journal of Pharmacology and Experimental Therapeutics* 273 (1995): 121–25.

Wason, Peter C. "Reasoning." In *New Horizons in Psychology*, edited by Brian M. Foss. Harmondsworth, UK: Penguin Books, 1966.

Wason, Peter C., and Philip N. Johnson-Laird. *Psychology of Reasoning: Structure and Content*. London: Batsford, 1972.

Watson, Gary. *Free Will*. Oxford: Oxford University Press, 1982.

Wegener, Alfred. *The Origin of Continents and Oceans*. New York: Dover, 1966.

Wegner, Daniel M. *The Illusion of Conscious Will*. Cambridge, MA: MIT Press, 2002.

Wegner, Daniel M., and Thalia Wheatley. "Apparent Mental Causation: Sources of the Experience of the Will." *American Psychologist* 54 (1999): 480–91.

Weintraub, Arlene. "Eyes Wide Open." *Business Week*, April 24, 2006.

Willoch, Frode, et al. "Phantom Limb Pain in the Human Brain: Unraveling Neural Circuitries of Phantom Limb Sensation Using Positron Emission Tomography." *Annals of Neurology* 48 (2000): 842–49.

Witkin, Herman A., et al. "Criminality in XYY and XXY Men." *Science* 193, no. 4253 (1976): 547–55.

Wright, Larry. "Argument and Deliberation: A Plea for Understanding." *Journal of Philosophy* 92, no. 11 (1995): 565–85.

———. *Better Reasoning: Techniques for Handling Argument, Evidence and Abstraction.* New York: Holt, Rinehart and Winston, 1982.

Yesavage, Jerome A., et al. "Donepezil and Flight Simulator Performance: Effects on Retention of Complex Skills." *Neurology* 59 (2002): 123–25.

INDEX